54 Structure and Bonding

Inorganic Elements in Biochemistry

With Contributions by
P. H. Connett H. Follmann M. Lammers
S. Mann J. D. Odom K. E. Wetterhahn

With 57 Figures and 22 Tables

Springer-Verlag
Berlin Heidelberg GmbH 1983

ISBN 978-3-662-15742-8 ISBN 978-3-540-44420-6 (eBook)
DOI 10.1007/978-3-540-44420-6

Library of Congress Catalog Card Number 67-11280

Originally published by Springer-Verlag Berlin Heidelberg New York Tokyo in 1983
Softcover reprint of the hardcover 1st edition 1983

2152/3140-543210

Table of Contents

Selenium Biochemistry
Chemical and Physical Studies

Jerome D. Odom

Department of Chemistry, University of South Carolina, Columbia, South Carolina 29208

A brief discussion of the vast and increasingly important area of organoselenium chemistry is presented as it relates to the biochemistry of this element. Organo selenoamino acids are the basic building blocks of many selenium-containing macromolecules, which are the primary focus of this article. First to be discussed are synthetic selenium analogues of sulfur-containing natural products such as selenocoenzyme A and selenobiotin. Secondly, are those enzymes which will indiscriminately incorporate selenium and the most important of these are considered. Finally, the most important and interesting class of selenium-containing macromolecules are the naturally occurring selenoproteins, of which there are eight currently known and only three which have been extensively studied. Physical studies which relate to or have the potential to aid in investigations of selenium biochemistry constitute the final topic.

Structure and Bonding 54

1 Introduction

Selenium, a group VIA element, exhibits many expected similarities in chemical properties to sulfur and tellurium[1-3]. Selenium has been classified as both a metal and a nonmetal and occupies a unique position in the Periodic Table. In its family it bridges between the nonmetals oxygen and sulfur and the metals tellurium and polonium, while in its period it lies between the metal (or metalloid) arsenic and the nonmetal bromine. Fortunately, for biochemical studies, the chemistry of selenium resembles that of sulfur more than any other element. Like sulfur, selenium displays a variety of valency states (-2, 0, $+2$, $+4$, $+6$) and the two elements have similar covalent radii and electronegativities[4]. Selenium has six naturally occuring isotopes, ^{74}Se (0.87%), ^{76}Se (9.02%), ^{77}Se (7.58%), ^{78}Se (23.52%), ^{80}Se (49.82%), ^{82}Se (9.19%) and several radioactive isotopes, the most important of which is ^{75}Se (a gamma emitter, $t_{1/2} = 120.4$ days) which is widely used as a tracer in biological studies and in certain radiologic diagnostic procedures. Of the naturally occurring isotopes only ^{77}Se has a nuclear spin ($I = ½$) and nuclear magnetic resonance (NMR) spectroscopy has recently proven very useful in the study of selenium compounds (see below).

Selenium is widely distributed in nature[5-7]. It ranks seventieth in order of abundance of the elements and comprises approximately 7×10^{-5} weight per cent of the earth's crust. Studies concerned with selenium content of minerals[8], meteorites and volcanic matter[9-13], soils[6, 14], plants[6, 8, 15-17], fossil fuels[16, 18, 19], water[20, 21] and air[16] have been reported.

Selenium is used industrially[7, 22] in the electronics industry in dry-plate rectifiers, xerography and photoelectric cells, in the metallurgic industry as a degasifier in stainless steel and in chromium plating solutions, in the glass and ceramics industry and in the manufacture of pigments. Additionally, selenium finds use as a catalyst in the preparation of some pharmaceuticals and is a constituent (in the form of selenium sulfide) of fungicides for the control of dandruff and dermatitis in humans and dermatitis, pruritis and mange in dogs. Other relatively minor uses of selenium occur in the explosives industry, in the rubber industry, in lubricants and in photographic photosensitizers and toners.

From a biological point of view, selenium is a paradoxical element, since it is highly toxic and yet it is an essential micronutrient for mammals, birds, several bacteria and probably for fish and many other animals species. Around 1930 attention was drawn to the toxic properties of selenium[6] which is quite prevalent despite its overall low concentration in the earth's crust. It was discovered that selenium, if present in the soil even in minute amounts, is taken up by vegetation and may cause damage to animals grazing on these plants. "Alkali disease" and "blind staggers", diseases of livestock known for some time[6], are now attributed to selenium poisoning. Alkali disease results from animals feeding on hay, grasses, etc. grown on soils high in selenium content and is a chronic form of poisoning, since lethal doses are not ingested. Blind staggers results from acutely high concentrations of selenium that occur when grazing animals eat selenium accumulator plants. Some of these plants are dependent on selenium for their normal growth and never occur in selenium-free soils. Although such "accumulator plants" are highly toxic to most animals, they are capable of supporting parasites, insect larvae with very high selenium tolerance. The relationship between selenium and vegetation still presents many interesting problems.

The occurrence and toxicology of selenium in foods has recently been reviewed[23]. Humans have an estimated dietary intake of selenium of from 4–35 μg/day in infants[24] to 60–250 μg/day in adults[25] with essentially all the selenium coming from cereals, grain, fish, meat and poultry. Although it is thought that selenium in plants is primarily in the form of selenoamino acids and in animals it is bound to proteins, toxicity studies with laboratory animals have primarily been carried out with the selenate (SeO_4^{-2}) and selenite (SeO_3^{-2}) ions and to a lesser degree with selenoamino acids. The degree of toxicity[7, 26] is $SeO_3^{-2} > SeO_4^{-2} >$ Se-amino acids and the ingested selenium is absorbed and most of it is subsequently excreted in the urine. A small portion in the form of methylated selenides is exhaled and excreted in perspiration[27, 28]. The selenium retained by the body is incorporated into proteins and is located primarily in the liver, kidneys and spleen[29].

Considerable debate exists in the literature regarding whether selenium is carcinogenic[30–40], and, in contrast, some workers contend that selenium has significant anticarcinogenic properties. In addition, selenium appears to exhibit "mild" teratogenic effects[30]. Epidemiologic evidence involves selenium in teratogenesis in pigs, sheep and cattle. Malformations among chickens and sheep raised in seleniferous areas have been reported. Birth of deformed infants and miscarriages in several pregnancies among female laboratory technicians exposed to selenium suggest the teratogenic potential of this element among humans.

In connection with the toxicity of selenium, it should be noted that arsenite presumably can overcome selenium poisoning and that selenium compounds have been used to treat heavy metal poisoning, arising from such metals as silver, cadmium, mercury and lead[6]. This is thought to occur by altering the coordination of these metals (e.g. selenium compounds have a greater affinity for mercury and methyl mercury than do the sulfur analogs) as well as their retention and distribution in the body.

With regard to the nutritional aspects of selenium metabolism, the essentiality of this element was established in studies which demonstrated that a number of deficiency diseases[6, 41] such as exudative diathesis in chicks and turkeys, necrotic liver degeneration in rats and swine[42] and white muscle disease[43], which is a nutritional type of muscular dystrophy often affecting young lambs and calves, could be overcome or prevented by administering small doses of selenium.

An increased awareness of the important role which selenium appears to play in nutrition has stimulated an increased effort to discover and more thoroughly understand the role of selenium in biochemical processes. This review will attempt to cover some aspects of what is currently known concerning the biochemistry of selenium.

2 Organic Chemistry of Selenium

Organoselenium chemistry is a vast and extensive area which has developed at an incredibly rapid pace in the last several years[44–46]. Of particular interest is that a number of useful functional group transformations can be achieved by the introduction and removal of selenium functions. Since this area has been thoroughly reviewed[47–51], this section merely endeavors to acquaint the reader with the general reactivities and names of organic selenium compounds, especially those which have or will potentially have biological interest. Table 1 lists formulas and names of some relevant selenium containing

Table 1. Relevant organic selenium compounds

Formula	Name
RSeH	Selenol
RSeX (X = Cl, Br)	Selenenyl Halide
RSe(O)OH	Organyl Seleninic Acid
$RSe(O)_2OH$	Organyl Selenonic Acid
R_2Se	Diorganyl Selenide
R_2Se_2	Diorganyl Diselenide
R_2SeX_2	Diorganyl Selenium Dihalide
R_2SeO	Diorganyl Selenoxide
R_2SeO_2	Diorganyl Selenone
R_3Se^+	Triorganyl Selenonium Ion
RC(O)SeH	Organyl Selenocarboxylic Acid
$RCH(SeR')_2$	Organyl Selenoacetal
$RR'C(SeR'')_2$	Diorganyl Selenoketal

organic molecules. In addition it should be noted that there are a large number of heterocyclic selenium compounds which have attracted the attention of researchers[52)].

In comparing organosulfur and organoselenium chemistry some useful generalizations can be made with regard to selenium in these systems:

(a) Usually organoselenium compounds are more reactive than their sulfur analogues presumably because of the slightly greater polarity and lower bond strength of the C-Se σ bond as well as other σ bonds such as the N-Se and O-Se bonds. For example reports in the literature document the fact that episelenides spontaneously eliminate selenium[53–55)] at about 25 °C, [1,3] and [2,3] sigmatropic rearrangements occur at much lower temperature with selenium[56–59)] and alkyl selenoxides undergo syn elemination by a factor of 10^3 faster than do sulfoxides[60, 61)].

(b) Nucleophilic attack at selenium occurs more readily in organoselenium compounds as is demonstrated by the fact that in general most organoselenium molecules are attacked at the selenium atom by organolithium reagents[62–65)]. Seebach has taken advantage of this by synthesizing α-lithio selenides via cleavage of selenoacetals and -ketals[66–68)]. The ease of nucleophilic attack at selenium was further demonstrated by Reich in a report that α-phenylseleno ketones rapidly deselenate in base[69)].

(c) Selenides and selenolate ions are more nucleophilic[70, 71)] and less basic[72)] than analogous sulfur molecules while selenoxides are more basic (and more polar) than sulfoxides[73)]. Selenoxides are also more stereochemically labile than sulfoxides and are thus easily racemized in neutral or weakly acidic media[74)]. With regard to oxidation-reduction reactions, it is easier to oxidize selenium than sulfur to oxidation state (IV) but more difficult to oxidize it to oxidation state (VI). The consequence of this is that selenides, selenols and hydrogen selenide are converted to selenoxides, seleninic acids and selenium dioxide, respectively, by oxidizing agents[51)]. To obtain selenones, selenonic acids and selenium trioxide is more difficult. In addition the selenol group (-SeH) is widely found in biological selenium compounds and unlike the thiol group (-SH), selenols are largely ionized and anionic at neutral pH, are better nucleophiles, are better leaving groups and have lower redox potentials compared to the analogous thiols[75)]. All of the above factors may potentially be important in the biochemistry of selenium.

3 Selenoamino Acids

Since selenium occurring in "accumulator" plants grown in selenium-containing soil was found to reside in plant proteins, research leading to the isolation and synthesis of selenoamino acids and their derivatives was stimulated. As pointed out by Ganther[76], the forms of selenium which are found in living systems are dependent upon the form and amount of selenium to which the organism is exposed as well as the species itself. Of the selenium compounds which have been found in living systems, the "low molecular weight" compounds are almost all selenoamino acids. (Other materials which have been reported or implicated are dimethyl selenide[77–79], dimethyl diselenide[80], trimethyl selenonium ion[81, 82] and even elemental selenium[83, 84]). These selenoamino acids are discussed in this section.

3.1 *Selenocystine*

This diselenide is largely found incorporated into proteins but the free selenoamino acid has been detected in plants after selenite has been administered. Selenocystine was also the first selenoamino acid synthesized[85, 86] although all early syntheses of this molecule suffered from low yields[85–90]. There are two methods which are currently used for the synthesis of this molecule. The method reported by Zdansky[91] and described in detail by Chu[92] yields DL- and meso-selenocystine in approximately 70% overall yield (Scheme 1).

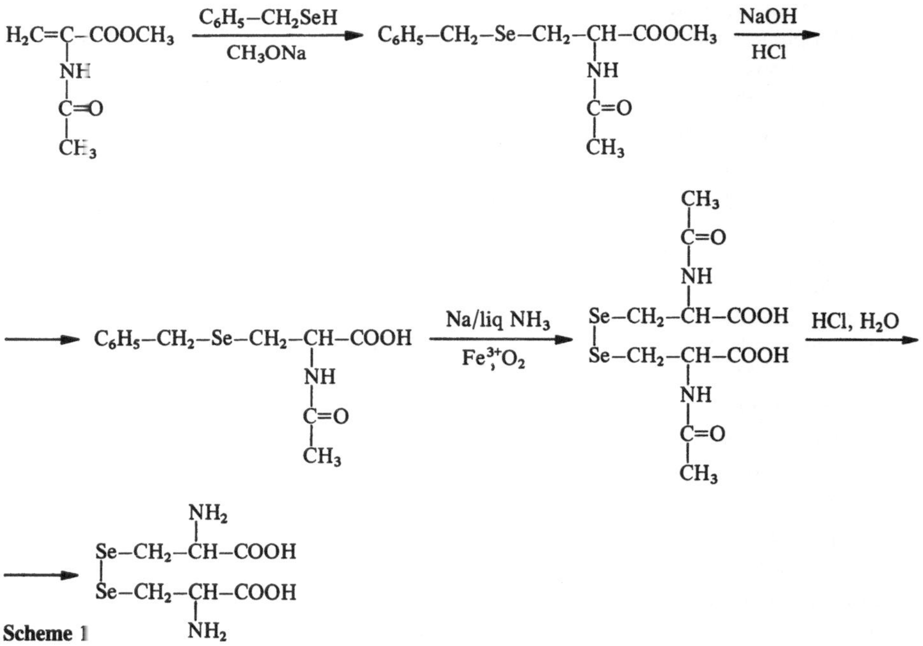

$H_2C{=}C(NHCOCH_3){-}COOCH_3 \xrightarrow[CH_3ONa]{C_6H_5{-}CH_2SeH} C_6H_5{-}CH_2{-}Se{-}CH_2{-}CH(NHCOCH_3){-}COOCH_3 \xrightarrow[HCl]{NaOH}$

$\longrightarrow C_6H_5{-}CH_2{-}Se{-}CH_2{-}CH(NHCOCH_3){-}COOH \xrightarrow[Fe^{3+}, O_2]{Na/liq\ NH_3} \begin{matrix} Se{-}CH_2{-}CH(NHCOCH_3){-}COOH \\ | \\ Se{-}CH_2{-}CH(NHCOCH_3){-}COOH \end{matrix} \xrightarrow{HCl,\ H_2O}$

$\longrightarrow \begin{matrix} Se{-}CH_2{-}CH(NH_2){-}COOH \\ | \\ Se{-}CH_2{-}CH(NH_2){-}COOH \end{matrix}$

Scheme 1

A second method[93–95] utilizes nucleophilic displacement of an O-p-toluene-sulfonate moiety of O-tosylated serine derivatives by either the benzyl selenolate anion or sodium hydrogen selenide. This latter method has been specifically investigated with regard to peptide synthesis with suitable protecting and assisting groups and although the yield is not as good (~ 50%), it is the preferred method for small scale production of optically active material.

Optically active selenocystine is soluble in water at 25° to the extent of 6.3×10^{-4} mol/l[86] while the solubility of DL-selenocystine in water at 25 °C and pH 7 is reported to be 2.35×10^{-3} mol/l[96]. Selenocystine and hexagonal cystine are assumed to be isomorphous with similar unit cell dimensions[97, 98]. Selenocystine is unstable in 6 M HCl at 110 °C with only 5% remaining after 6 h[96] although, interestingly, 60% of selenocystine in peptides remains intact under these same conditions[99]. In an interesting chemical comparison, Caldwell and Tappel[100] found that reduction of hydroperoxides with selenocystine occurred with greater ease and was more extensive than with cystine. It was also reported that selenocystine reacts more slowly with organic hydroperoxides than with hydrogen peroxide itself.

3.2 Selenocysteine

The parent selenol, $HSeCH_2CH(NH_2)COOH$, rapidly oxidizes in air and the diselenide, selenocystine, is obtained. Solutions of selenocysteine can be prepared by reduction of selenocystine[101], by debenzylation of Se-benzyl-selenocysteine[102, 103] and by nucleophilic displacement of O-tosylated serine derivatives by $C_6H_5CH_2Se^-Na^+$ or NaHSe[95, 104]. A number of Se-alkyl[103] and Se-aryl[85, 87, 91, 94] derivatives of selenocysteine have been reported with the most important from a biological point of view being Se-methylselenocysteine, $CH_3SeCH_2CH(NH_2)COOH$, which is the predominant soluble form of selenium in accumulator plants of the genus *Astragalus*. It is a crystalline solid which was first isolated in 1960 by chromatography[105] and its biosynthesis from selenite was demonstrated several years later[106, 107].

A very interesting aspect of biological selenium compounds is the occurrence of selenocysteine in a protein component (selenoprotein A) of glycine reductase, in formate dehydrogenase and in glutathione peroxidase (see below). As pointed out by Stadtman[108], the incorporation of this rather unusual amino acid into proteins in a very specific way poses something of a mystery and at present the mechanism by which this occurs is not known.

3.3 Selenocystathionine

This particular amino acid ($HOOCCH(NH_2)CH_2Se(CH_2)_2CH(NH_2)COOH$) has been isolated from several types of plants. It was first discovered in the *Astragalus* plant by Horn and Jones[109]. The selenium accumulators *S. pinnata*[110] and *Neptunia amplexicaulis*[111] also biosynthesize selenocystathionine from selenite. In addition, this amino acid is the cytotoxic compound in monkey nut seeds[112, 113].

The synthesis of optically active selenocystathionines in the laboratory has been accomplished by Zdansky and Pande[114, 115].

3.4 Selenohomocystine and Selenohomocysteine

DL-selenohomocystine, $(HOOCCH(NH_2)(CH_2)_2Se)_2$, was first synthesized by Painter[116] by treating ethyl 2-benzamido-4-chlorobutyrate with NaHSe or $C_6H_5CH_2Se^-Na^+$. The benzyl selenolate reaction afforded good yields of both selenohomocystine and Se-benzylselenohomocysteine. Since that time several other synthetic methods have been reported[115, 117–120] primarily for the synthesis of aryl derivatives of the selenol, selenohomocysteine, which can then be easily oxidized to the diselenide. In addition Zdansky[121] has reported the synthesis of a series of methyl-substituted Se-benzyl-selenohomocysteines which contain the methyl groups on carbon atoms 2, 3 or 4. Some di- and trimethylselenohomocysteines have also been prepared[121] which were then debenzylated with sodium in liquid ammonia and oxidized in air in the presence of Fe^{+3} ions to yield the corresponding methylated selenohomocystines[121]. Selenohomocystine occurs naturally from the metabolism of ^{75}Se-selenomethionine in the plant *Astragalus crotalariae*[122].

3.5 Selenomethionine

As with selenocystine, the amino acid selenomethionine, $CH_3Se(CH_2)_2CH(NH_2)COOH$, is largely incorporated into proteins but has been found in the free form to some extent. Onions (*Allium cepa*) which had been injected with radioactive selenite were reported to contain selenomethionine[103] as were ethanol extracts of red clover, white clover and rye grass[84]

DL-Selenomethionine was initially synthesized by Painter[116] via a sodium/liquid ammonia reduction of DL-selenohomocystine followed by an alkylation of the resulting sodium selenohomocysteinate with methyl iodide. Other syntheses have since been reported[123, 124] including synthetic pathways for the preparation of optically active material[115, 119, 120, 125] and isotopically labeled material[126–128]. DL-Selenomethionine has a solubility in water at 30° and pH 7 of 0.108 M[129] which is considerably less than that for L-methionine (0.386 M). After seven hours of hydrolysis[129] under anaerobic conditions in 6 N HCl at 110 °C selenomethionine is completely decomposed (under the same conditions 95% of methionine remains). Chemically, selenomethionine appears to be more reactive than methionine[129]. For example, with cyanogen bromide, selenomethionine completely reacts in 0.1 M HCl in fifteen minutes while methionine requires twenty-four hours for the same reaction. In both cases the end product is homoserine. Although not as marked, this difference in reactivity was also confirmed in the reaction with hydrogen peroxide[129].

Interestingly Se-methylselenomethionine, $(CH_3)_2Se^+(CH_2)_2CH(NH_2)COOH$, has been identified as the predominant soluble selenium compound which is synthesized from selenite by species of *Astragalus* which lack the ability to accumulate selenium[130]. As mentioned previously, the predominant selenium compound in the accumulators is Se-methylselenocysteine.

4 Selenium-Containing Biological Macromolecules

In the field of selenium chemistry, the most exciting developments in the past several years have been discoveries in the area of selenium biochemistry, specifically selenium-dependent biochemical processes. This area has been excellently reviewed by Stadtman in several articles[75, 108, 131, 132]. It was not until 1957 that evidence was presented with regard to the essential nature of selenium[133] although its toxic nature had been recognized for some time. By an interesting coincidence, in the same year properties of glutathione peroxidase, a newly detected enzyme, were reported by Mills[134]. Not until 1972 was the relationship between selenium and this enzyme reported[135, 136]. Currently selenium is known to be a normal component of several enzymes, proteins and some aminoacyl transfer nucleic acids (tRNAs). In these materials the selenium is obviously found in a precisely specified location. In addition, because of the many similarities between sulfur and selenium, there are a number of sulfur-containing biological molecules which can utilize selenium instead. These include chemically synthesized selenium analogs of a variety of naturally occurring sulfur-containing molecules as well as enzymes that utilize both sulfur and selenium compounds as substrates. This section will cover all three of these types of systems.

4.1 Synthetic Selenium Analogues of Sulfur-Containing Natural Products

As discussed in a previous section, organoselenium chemistry has recently evolved into an area which is rich and varied. Synthetic organic chemists have been instrumental in stimulating the interest in research concerning selenium analogues of natural products which contain sulfur. The biological activity of many of these molecules has also been an area of growing interest.

Coenzyme A (CoA) plays an extremely important role in metabolic pathways. The function of Coenzyme A is to serve as a carrier of acyl groups in enzymatic reactions involved in fatty acid oxidation, fatty acid synthesis, pyruvate oxidation and biological acetylations[137]. The chemical mechanism by which CoA carries acyl groups was established by Lynen[138, 139] demonstrating the vital importance of the single thiol group in the molecule. Gunther and Mautner, recognizing that although the similar sizes of sulfur and selenium should not have an appreciable effect on the ability of the molecule to fit enzymic receptor sites but electron distributions and thus reactivity might well vary, reported the synthesis and activity studies of selenocoenzyme A in 1965[140]. Selenocoenzyme A had neither catalytic nor inhibitory activity in the phosphotransacetylase system while it did exhibit catalytic activity in an ATP acetylation of 4-aminoazobenzene by a bicarbonate extract of pigeon liver acetone powder[140].

A second well-known sulfur-containing coenzyme is biotin[141–143]. Biotin, a member of the B vitamin complex, plays an essential nutritional role in carboxylation reactions. It is the coenzyme of carboxylases and catalyzes fundamental metabolic processes. Its relative unavailability from natural sources spurred great interest in synthetic approaches, the most recent of which appeared in 1977[144, 145]. The synthesis of selenobiotin was published in 1976 by Bory and Marquet[148]. The biological activity of selenobiotin as a growth factor for biotin-requiring organisms is very similar to biotin itself[147, 148]. Interestingly, initial evidence has recently been presented suggesting that selenobiotin is a naturally occurring and functional biomolecule[149].

In certain ferredoxins (2 Fe-2 S) substitution of selenium for all or part of the sulfur has been accomplished and the proteins were found to be biologically active[150–152]. Another notable example involved the preparation of selenated tRNA by transformation of the 4-thiouridine moiety into 4-selenouridine[153]. This was accomplished by treatment of the tRNA(-SH) with cyanogen bromide followed by conversion of the resulting tRNA-(-SCN) to tRNA(-SeH) with NaHSe. This allows the production of significant amounts of seleno-tRNA but biological effects have not yet been reported.

Finally, it should be mentioned that there are numerous examples in which oxygen or methylene groups of selected biomolecules are replaced by selenium including nicotinamide-adenine dinucleotide phosphate (NADP) in which selenium replaces the oxygen in the amide group of nicotinamide[154], β-selenaproline and γ-selenaproline in which methylene groups are replaced by selenium[155, 156] and lastly, selenomethotrexate in which methylselenocysteine replaces the glutamic acid moiety[157].

4.2 Enzymes Which Catalyze Incorporation of Selenium As Well As Sulfur

There are a number of examples of enzymes which under normal conditions catalyze reactions that result in the transformation of a particular sulfur compound which have also been found to undergo the same reaction with the analogous selenium compound. This section will point out some of the more important examples but will not attempt to be exhaustive since the topic has been extensively discussed elsewhere[131, 132, 158].

The replacement of sulfur-containing amino acids in proteins by their selenium analogs has received some attention. For example, a variant of *Escherichia coli* was shown to indiscriminately incorporate selenium into its proteins[159, 160]. In fact, the β-galactosidase isolated from this system, which had 70–75% of its methionine residues replaced by the selenium analog, exhibited a virtually unaltered catalytic activity. Similarly, the indiscriminate incorporation of selenomethionine in proteins of the rat[132] as well as *E. coli* suggests that the enzymes methionyl-tRNA synthetase and amino acid polymerase can accept both methionine and selenomethionine and their corresponding tRNA derivatives. There is some evidence which suggests that a similar phenomenon occurs with cysteine and selenocysteine[161, 162].

Selenium-containing biomolecules can also be prepared *in vitro* by enzymatic methods. For example, tRNA sulfur transferase catalyzes the formation of tRNA-containing 4-thiouracil in an ATP-dependent reaction using cysteine as the source of sulfur. It has been reported that selenocysteine can function as a substrate in this reaction with the result being the synthesis of tRNA containing 4-selenouracil[132, 163, 164]. Other notable enzyme catalyzed reactions in which selenium-containing substrates have been shown to react in place of the corresponding sulfur-containing substrates are adenosine-5′-selenophosphate (ATP sulfurylase)[165], selenocysteine (cysteinyl-tRNA synthetase) and Se-adenosylmethionine (S-adenosylmethionine methyl transferase)[166].

With regard to these enzymes, it has been suggested[132] that their lack of specificity involved in sulfur metabolism may play an integral role in the toxicity of selenium compounds. At this time definitive results have not been obtained regarding the biological effects of metabolized selenium by one or more of these nonspecific enzymes, but it would certainly seem logical that large scale and indiscriminate incorporation of selenium for sulfur in key biological macromolecules could be harmful. This particular area deserves much more study with regard to selenium toxicity.

4.3 Naturally Occurring Selenoproteins

This is an area which has only very recently attracted intensive study. In fact, as recently pointed out by Stadtman[108)], detailed studies of the specific biochemical role of selenium at the enzyme level dates only from 1972. However, research is progressing rapidly. To put this into perspective, a 1979 report[137)] stated that "there are at least three, and possibly four", selenoproteins. A 1980 article[75)] listed eight known selenoproteins! Table 2 lists these selenoproteins and their source. Interestingly, the form of selenium which has been determined in all animal and bacterial proteins examined thus far appears to be selenocysteine. Both bacterial and mammalian enzymes apparently are involved in redox reactions (an exception may be thiolase; see below). Selenomethionine, as mentioned previously, is a predominant form of selenium in many grains and grasses, but it has not been detected in mammalian enzymes.

It is important to remember that only minute amounts of selenium are required for the synthesis of these proteins and whatever mechanism(s) accounts for the incorporation of the selenium, there must be an extremely effective discrimination between it and sulfur. Otherwise sulfur, being so much more abundant, in any kind of competitive mechanism, would completely overwhelm the selenium incorporation. In this regard it is interesting to note that even when bacteria are cultured in the presence of a tremendous molar excess of sulfur, clostridial glycine reductase is synthesized normally, containing the single selenocysteine in the same polypeptide chain as two cysteine and three methionine residues[167, 168)].

The mechanism of incorporation of selenium, particularly into mammalian proteins, in a relatively controversial issue and this is an area which deserves, and will undoubtedly

Table 2. Naturally occuring selenoproteins

Selenoprotein	Source
Mammalian	
Glutathione Peroxidase	Mammals and birds
Selenoprotein (10,000 daltons)	Lambs
Selenoprotein (15,000 daltons)	Rat testes
Bacterial	
Formate Dehydrogenase	*E. coli* and anaerobic bacteria
Glycine Reductase	*Clostridium sticklandii*
Nicotine Acid Hydroxylase	*Clostridium barkeri*
Thiolase	*Clostridium kluyveri*
Xanthine Dehydrogenase	*Clostridium acidiurici* and *Clostridium cylindrosporum*

receive, a great deal of attention in the future. The only selenoamino acid which has been detected in proteins thus far is selenocysteine and one suggested mechanism[169] involves the *in situ* formation of selenocysteine from cysteine or serine and selenide, similar to the production of cysteine from serine and sulfide in the cysteine synthase reaction. However, two experimental studies are currently in conflict with this mechanism[170, 171]. Whatever mechanism is postulated there are several interesting questions which must be addressed. The absence thus far of selenomethionine in mammalian tissues poses an interesting question. Is this amino acid converted to selenocysteine? A recent report[172] that selenomethionine is just as effective as the selenite ion for the stimulation of glutathione peroxidase activity in selenium depleted rats lends affirmative evidence to this question. A second question which arises is if it is only coincidental that in both the mammalian enzyme glutathione peroxidase and the bacterial enzyme glycine reductase that each subunit contains the essential selenocysteine, two cysteines and three methionines? The answers to these and other questions relating to this incorporation mechanism will hopefully be answered in the near future.

4.3.1 Glutathione Peroxidase

Before glutathione peroxidase was known to be a selenoprotein[135, 136, 173], glutathione peroxidase had been characterized kinetically and mechanistically[174, 175] and had even been isolated in crystalline form[173, 176]. The function of this enzyme is to catalyze the decomposition of peroxides (H_2O_2 and a variety of organic peroxides) which can damage red blood cell membranes and other tissues which function in aerobic surroundings (Eq. 1)[177, 178].

$$2\,GSH + ROOH \rightarrow GSSG + ROH + HOH \quad (R = H, \text{organic moiety}) \tag{1}$$

Glutathione peroxidase has a molecular weight between 76,000 and 92,000 daltons and contains four identical 19–23,000 dalton subunits[179–182], each of which contains a single selenocysteine residue[183, 184]. Purified forms of the enzyme have been obtained from the red blood cells of cattle[179], sheep[180] and humans[181] as well as from rat liver[182] and bovine lens[185]. An amino acid composition of homogeneous rat liver glutathione peroxidase has been determined[182] and an X-ray diffraction study at 2.8 Å resolution has been published[184]. There is some discrepancy between these two studies since the amino acid analysis indicates two cysteine residues per subunit which were not detected by the X-ray study. This point will have to be clarified. The X-ray study has shown that the active selenocysteines are found in flat depressions on the surface of the molecule which exists in long α-helices.

Debate still exists concerning the form of the selenocysteine residues in the reduced from of the enzyme (SeH?). It is not clear whether there is a selenium valency change upon reaction or if, during the reaction with peroxide substrate, the selenium in the enzyme is converted to oxygen-containing derivatives (e.g. selenenic acids, Enz-RSeOH). Direct experimental evidence concerning this oxidation step is needed.

4.3.2 Muscle Selenoprotein

This selenoprotein is not nearly so well characterized as glutathione peroxidase and its biochemical role is still unknown. This is a small molecular weight selenoprotein (10,000 daltons) which was reported in 1972[186]. It was isolated from heart and muscle homogenates of normal lambs. This protein is missing in animals which suffer from white muscle disease, which is a type of muscular dystrophy resulting from a deficiency of selenium in the diet[187, 188]. Further work on this selenoprotein has been reported[189] and recently the form of selenium present was identified as selenocysteine[170, 190].

4.3.3 Selenoprotein From Rat Testes

Very recently, a selenium-containing protein with a molecular weight of approximately 15,000 daltons was reported by McConnell[191]. The protein was isolated from the testes of sexually mature rats and apparently does not appear until the onset of sexual maturity. Calvin has also recently reported[192] a selenoprotein which he isolated from rat sperm tail and it will be interesting to see if these two proteins are identical. It has been felt for some time that selenium was necessary in the diet of domestic animals, especially male animals, to ensure fertility and efficient reproduction. The isolation of the above selenoprotein(s) may have some bearing on this aspect of fertility.

4.3.4 Formate Dehydrogenases

This group of bacterial enzymes is an interesting collection which have been identified in *E. coli*[192] and several anaerobic bacteria, among them being *Clostridium thermoaceticum*[193], *Clostridium acidiurici*[194], *Clostridium cylindrosporum*[194], *Clostridium formicoaceticum*[195, 196], *C. stricklandii*[197] and *Methanococcus vannielii*[198]. However there are also selenium independent formate dehydrogenases which have been reported[108, 199, 200].

The active form of selenium in *M. vannielii* has been identified as occurring as selenocysteine residues[198]. In addition to selenium, the known selenium-dependent formate dehydrogenases contain molybdenum and iron-sulfur centers. The *E. coli* enzyme also contains cytochrome *b* subunits[192]. It has been characterized as an approximately 600,000 dalton protein which contains three different types of subunits (a $\alpha_4\beta_4\gamma_2$(or γ_4) structure). The α subunits which are 110,000 dalton subunits contain the 4 gram atoms of selenium (presumably selenocysteine)[192]. Formate dehydrogenases are rapidly deactivated by oxygen and have proven difficult to isolate in pure, catalytically active form. Because of these problems, neither the exact composition nor the catalytic roles of selenium, molybdenum or the iron-sulfur centers are known at this time.

4.3.5 Glycine Reductase

Clostridal glycine reductase is the bacterial enzyme which has been investigated most thoroughly[108, 132]. This enzyme catalyzes the reductive deamination of glycine to ammonia and acetate with the concomitant synthesis of ATP (Eq. 2). As has been

$$\mathrm{H_2C(COOH)(NH_2)} + \mathrm{R(SH)_2} + \mathrm{ADP} + \mathrm{P_i} \longrightarrow \mathrm{NH_3} + \mathrm{CH_3COOH} + \mathrm{R{<}S_2} + \mathrm{ATP} \qquad (2)$$

pointed out by Stadtman the reduction reaction is metabolically important because it serves as an electron sink for a number of amino acid fermenting clostridia[132]. Stadtman has been instrumental in the characterization of this selenoprotein and compositional studies of the *C. sticklandii* glycine reductase have shown that the complex consists of three components[167, 201]. The selenium is contained in selenoprotein A (12,000 daltons) which is acidic and stable to heat[167, 168, 202, 203–206]. Protein B (200,000 daltons) is reported to contain at least one carbonyl group which is essential for its activity[108] and the fraction C protein (250,000 daltons) has been shown to co-purify with iron[201]. Selenoprotein A contains one gram atom of selenium per mole in the form of a selenocysteine residue[68, 203]. In fact this was the first selenium-dependent protein in which the active form of selenium was identified. As pointed out earlier in this section this selenoprotein (in the A component) also contains two cysteine residues and three methionine residues as does glutathione peroxidase which perhaps argues for post-translational modification as a mechanism for the incorporation of selenium into these selenium-dependent systems. The acidity of the selenoprotein A fraction is thought to be due to a high content of aspartate and glutamate residues. Although fraction C has not been obtained in homogeneous form, its co-purification with iron as well as other similarities have led to a comparison of glycine reductase with another iron-containing enzyme complex, *E. coli* ribonucleotide reductase[108].

4.3.6 Nicotinic Acid Hydroxylase

While it has not yet been definitively determined that selenium is essential in the enzyme, this appears very likely. Nicotinic acid hydroxylase is the catalyst in the first step of a series of anaerobic reactions which result overall in the fermentation of nicotinic acid to acetate, ammonia, carbon dioxide and propionate[204, 205]. The first reaction is shown below and is the addition of water to nicotinic acid to yield 6-oxonicotinic acid and to generate NADPH[206, 207] (Eq. 3).

$$\text{Nicotinic acid (COOH)} + \mathrm{NADP^+} + \mathrm{H_2O} \rightleftharpoons \text{6-oxonicotinic acid (COOH, O=)} + \mathrm{NADPH} + \mathrm{H^+} \qquad (3)$$

Recent studies demonstrated that *C. barkeri* produces increased levels of nicotinic acid hydroxylase when cultured in selenium rich media and that when the enzyme was purified from *C. barkeri* cells labeled with ^{75}Se-selenite, the ^{75}Se and the enzyme activity co-purified[208]. Further experiments are required to establish that selenium is indeed required by this enzyme and to identify the form of selenium present.

4.3.7 *Thiolase*

This enzyme is the most recently discovered selenium dependent enzyme[209] and this may be the first example of a selenoenzyme that does not function as a redox catalyst. Thiolase catalyzes the CoA-dependent cleavage of acetoacetyl-CoA to yield 2-acetyl-CoA (Eq. 4).

$$\text{Acetoacetyl-CoA} + \text{CoA-SH} \leftrightarrows \text{2-acetyl-CoA} \qquad (4)$$

Interestingly, Goldman[210] reported that in the anaerobic organism *Clostridium kluyveri,* the reaction was catalyzed in the direction of acetoacetyl-CoA. This can be understood with the discovery in Stadtman's laboratory that a second thiolase in this particular organism contains selenium since the reverse reaction could occur more easily if an enzyme-bound selenolacyl ester instead of a thiolacyl ester was an intermediate. As stated previously, a selenol is generally a better leaving group that a thiol which would facilitate the condensation reaction. Studies now are needed to compare these two thiolases from the same organism, only one of which contains selenium.

4.3.8 *Xanthine Dehydrogenase*

This enzyme, as well as nicotinic acid hydroxylase[208], was recently reported by Andreesan[211] to be a selenoenzyme. The discovery of both these enzymes was based on the clever assumption that selenium might well be a component of multisubunit enzymes containing redox centers such as iron-sulfur, flavin, molybdenum, etc. When *Clostridium acidiurici* was cultured in media with supplemental selenium, an elevated activity of xanthine dehydrogenase was observed. The clostridial enzyme[212] is comparable to mammalian xanthine oxidases in that it contains flavin adeninedinucleotide (FAD), molybdenum and nonheme iron. This enzyme functions *in vivo* under anaerobic conditions and appears to catalyze the reduction of uric acid to xanthine. Again it will be interesting to learn the form of selenium in this enzyme.

5 Spectroscopic Properties of Selenium Compounds

With regard to biochemical selenium compounds, the whole area of spectroscopic studies is sorely lacking. However, due to very nature of spectroscopy and the complexity of most selenium-containing biological systems, these studies must necessarily be initiated with a thorough understanding of the spectroscopic properties of model systems. As might be expected small inorganic and organic selenium-containing molecules have received the lion's share of attention from spectroscopists. In this section no attempt will be made to be comprehensive; rather, a general, overall view of several spectroscopic areas and studies which are potentially relevant to selenium biochemistry will be cited. Selenium-77 nuclear magnetic resonance (NMR) spectroscopy potentially offers some very exciting possibilities for future research with selenium macromolecules and this area will be described in some detail. In addition, the reader is referred to Ref. 45 which does an excellent job of supplying references to infrared, nmr, mass and UV-visible spectroscopic studies of organoselenium compounds.

5.1 Infrared and Raman Spectroscopy

Vibrational spectroscopy of organoselenium compounds is a relatively "young" area and most reports have been published since approximately 1965[213]. Raman, instead of infrared, spectroscopy may well prove to be the better of the two techniques since, unlike bonds to oxygen, bonds to selenium do not usually give rise to intense absorptions in infrared spectroscopy. In addition, many absorptions associated with selenium modes are in the region 1100–600 cm^{-1} which normally is a "busy" region which may make assignments difficult. Deformational modes of many organoselenium compounds are below 400 cm^{-1} which is not such a crowded region and this spectral area may be more valuable analytically.

As expected, when comparing vibrational spectra of sulfur and selenium compounds, the absorptions due to bonds to selenium occur at lower frequencies than the corresponding sulfur modes. This frequency shift is normally 50–150 cm^{-1} but may be more for selenium-oxygen compounds when compared to sulfur-oxygen compounds[213].

Selenols are of particular interest with regard to biochemical selenium compounds and the Se-H stretch normally appears from 2280–2330 cm^{-1} [214–217]. This band is not influenced significantly by concentration effects, solvent effects or the state of aggregation, demonstrating that, as expected, selenols have very little tendency to form intermolecular hydrogen bonds[214].

Other important compounds include selenides and diselenides. The Se-C stretch in compounds of the general formula R-Se-R (R=alkyl) is 550–610 cm^{-1} [218–223] with a band at ca. 590 cm^{-1} being very diagnostic[223] of ν(C-Se). In addition, the intensity of this band appears to be dependent on the formal charge of the selenium atom[224]. The Raman spectra of selenomethionine have recently been studied under a variety of conditions of pH and oxidizing potential[225]. An intense, polarized Raman band at 600 cm^{-1} has been assigned to ν(Se-C) in this selenoamino acid. In dialkyl diselenides the Se-Se stretch occurs over the range 286–293 cm^{-1} [226, 227] and the C-Se stretch in these same compounds appears in the range 507–570 cm^{-1} [228] although, in general, ν(C-Se) occurs in the same region as in selenides. Thus, in a recent study of selenocystine[225], again under a variety of conditions, the Raman spectrum in HCl exhibited polarized peaks at 598 cm^{-1}, assigned to the Se-C stretch, and 288 cm^{-1}, assigned to the Se-Se stretch.

Finally, vibrational absorption bands of selenium-oxygen compounds might be of interest in the study of selenium in biological systems and the basic classes of Se-O molecules have been exhaustively studied[213]. Relevant bands are 1010–1040 cm^{-1} (ν_{as} (O-Se-O) in selenates), 930–960 cm^{-1} (ν_s (O-Se-O) in selenates), 930 cm^{-1} (ν (Se=O) in selenites), 850–900 cm^{-1} (ν (Se=O) in selenenic acids, esters and anhydrides), 800–840 cm^{-1} (ν (Se=O) in selenoxides) and 680–700 cm^{-1} (ν (Se-OH) in selenenic acids)[213].

At this time it would appear prudent for a vibrational spectroscopist to undertake a careful and detailed experimental vibrational study of solutions of selenocysteine under various conditions since selenocysteine residues seem to occur in most, if not all, naturally occurring selenoproteins. The facile oxidation of this selenoamino acid must be kept in mind and any vibrational study must be carried out under carefully controlled conditions.

5.2 Ultraviolet-Visible Spectroscopy

Aliphatic selenols do not have characteristic absorption maxima although the aliphatic selenolate ion absorbs in the range 243–253 nm with a rather intense absorption, log $\varepsilon \sim 3.7$[229]. This is about 10 nm higher than the corresponding band observed for aliphatic thiolate ions (230–240 nm)[230]. Again, because of the facile oxidation of selenols to diselenides, UV-visible spectra of these molecules must be obtained under carefully controlled experimental conditions and results must be interpreted cautiously. This was clearly demonstrated when the spectrum of selenophenol[231] was later shown to be a spectrum of a mixture of selenophenol and diphenyl diselenide[232].

Very little work has been reported for alkyl selenides. Diethyl selenide has a maximum at 250 nm with log $\varepsilon = 1.70$[233]. As expected more research has been reported for phenyl alkyl selenides[234] which are really not of concern with respect to biological systems. However, in diselenides, the electron-rich Se-Se bond is a distinct chromophore which was obvious in 1936 when Fredga reported[235] that diselenides of the general formula $HO_2CCH_2Se\text{-}(CH_2)_n\text{-}SeCH_2CO_2H$ were yellow when no methylene groups were present between the selenium atoms (i.e. n = 0) and were colorless when methylene groups were present (i.e. n $\neq$ 0). The nature of this absorption in diselenides (and disulfides) has been discussed[236, 237, 238] in terms of molecular orbital theory in which the bonds formed by selenium are formed by essentially unhybridized p orbitals. Thus the two unshared pairs on each selenium atom are contained in an *s* and a *p* orbital. The transition which is observed is presumably a $\pi^* \rightarrow \sigma^*$ transition whose energy is dependent on the dihedral angle of the diselenide which in open-chain diselenides is approximately 90°. These diselenides exhibit absorption maxima at approximately 312 nm[239]. Electron-withdrawing substituents increase the energy of the lowest electron transition[237, 240] and hyperconjugative effects have been proposed to cause a shift in the opposite direction to lower energy[236, 241]. Also, of possible importance is the fact that saturated, aliphatic selenenyl sulfides, R-S-Se-R, exhibit only one ultraviolet absorption demonstrating that the S-Se moiety acts as a single chromophore[242].

5.3 Chiroptical Studies

The term "chiroptical" refers to spectroscopic properties (primarily the techniques of optical rotatory dispersion (o.r.d.) and circular dichroism (c.d.)) which are dependent upon the chirality of the compound under investigation. A few relevant studies have appeared in this area and will be briefly discussed. The area has been extensively reviewed for organoselenium compounds[243].

Diselenides, like their oxygen and sulfur anologues, are chiral molecules. However because of the relatively long Se-Se bond and the low barrier to rotation about this bond the molecules in general do not exist in a single configuration unless constrained in a ring system. This was amply demonstrated in an o.r.d. study[244, 245] of a cyclic and several non-cyclic diselenides including selenocystine. The cyclic diselenide exhibits an amplitude of the Cotton effect at the longest wavelengths in the o.r.d. curve of a = + 87 whereas the non-cyclic diselenides exhibit essentially no Cotton effect. Synthetic studies have accomplished the replacement of the disulfide bridge in diaminooxytocin with both possible S-Se groups as well as a Se-Se bridge[246–248]. A c.d. study[249] of these compounds revealed that the bands assigned to the Se-Se chromophore increased in intensity with temperature which was attributed to a change in torsional angle.

A recent circular dichroism study reported the chiroptical properties of four selenoamino acids all of which are selenides[250]. The compounds studied were optically active selenocystathionine and its allo diastereomer, selenomethione and selenolanthionine, $Se(CH_2CH(NH_2)COOH)_2$. In the 190–250 nm region, positive Cotton effects of the carboxyl and selenide chromophores were found to correlate with L(=S) or L,L(=S,S) absolute configurations.

In addition the C-Se-C chromophore appears to have three optically active transitions at approximately 225, 210 and 195–200 nm[250]. Finally it should be mentioned that a number of selenophenyl and selenonaphthyl esters of L-amino acids have been studied and a band appearing at about 225–245 nm in some of these compounds was attributed to a transition involving the unshared electron pairs on the Se atom[251].

5.4 X-Ray Diffraction Studies

X-ray diffraction studies of biological molecules are becoming more and more commonplace and this technique is expected to be a valuable tool in elucidating structural parameters of biological selenium systems. In fact as mentioned previously, a x-ray crystallographic study of glutathione peroxidase at 2.8 Å resolution has recently been reported[184]. This is the only selenium-containing macromolecule for which a structure has been determined but smaller organoselenium molecules have been studied[252] and some general features will be discussed.

In aliphatic organoselenium compounds the selenium-carbon bond length is in the range 1.95–1.99 Å while carbons in an aromatic system form bonds to selenium which have an average length of about 1.93 Å[252]. The carbon-selenium double bond (C=Se) appears to have a length of 1.82–1.87 Å except in 2,4-diselenouracil where the bond lengths are 1.89 and 1.99 Å[253]. The bond angle formed by selenium and two carbon atoms can also generally be classified. The angle is less than 100° when selenium is bonded to two non-cyclic aliphatic carbon atoms and when the carbons are contained in an aromatic ring the angle is about 107°C.

In diselenides, the Se-Se bond length appears to be relatively constant. For example, one can compare the Se-Se bond lengths in dimethyl diselenide (2.325 Å)[254], bis-(diphenylmethyl)-diselenide (2.285 Å)[255], diphenyl diselenide (2.29 Å)[256] and 4,4′-(dichlorodiphenyl)diselenide (2.33 Å)[257]. In these same compounds the dihedral angle is 87.5°, 82°, 82 ± 3° and 74.5°, respectively. In a crystal structure[258] of selenium dibenzenesulfinate, $C_6H_5S(O)_2SeS(O)_2C_6H_5$, the Se-S distance (2.20 Å) as expected, is shorter than reported Se-Se distances.

As mentioned in the infrared and Raman section, vibrational spectroscopy shows little or no tendency for selenols to form intermolecular hydrogen bonds. In an interesting comparison of physical techniques (although with different compounds), x-ray studies of selenourea[259] and a derivative[260] as well as 2,4-diselenouracil[253] exhibit evidence for Se $\cdots$ HN hydrogen bonds.

5.5 Nuclear Magnetic Resonance Spectroscopy

Until very recently, the only nucleus which had been extensively studied in organoselenium compounds was the proton. During the 1960's a vast number of proton studies were reported and these studies have been tabulated and reviewed[261]. However, even though some excellent and interesting investigations have appeared, proton studies have not been extremely helpful due to the rather limited chemical shift range of this nucleus in organic molecules.

The nucleus which offers the most exciting possibilities is selenium-77; however, it was not until the late 1970's that investigations with this nucleus began appearing with any regularity. In fact, a 1978 book entitled *NMR and the Periodic Table*[262] devoted only ten pages to ^{77}Se NMR and only five papers were referenced[263–267] which had utilized pulsed Fourier transform (FT) methods. In fact, the development of pulsed FT NMR spectroscopy has greatly facilitated the observation of many of the so-called "less common" nuclei, particularly in biological systems, and has aided in the structural elucidation of many proteins and enzymes[268–270]. Since many of these molecules contain essential sulfhydryl groups, ^{33}S NMR spectroscopy could become an extremely powerful physical tool. However, the ^{33}S nucleus is quadrupolar ($I = 3/2$) and suffers from very low natural abundance (0.74%) and relatively low sensitivity (2.26×10^{-3} with respect to the proton). Fortunately, the chemistry of sulfur and selenium are sufficiently similar that selenium compounds retain the activity demonstrated by their sulfur cogeners. The selenium-77 nucleus is very amenable to the pulsed FT NMR experiment. It offers adequate sensitivity (6.93×10^{-3} with respect to the proton and 2.98 compared to ^{13}C) and natural abundance (7.5%) as well as being a spin-½ nucleus. Additionally it has been demonstrated that selenium-77 possesses a large chemical shift range (~3000 ppm) and is extremely sensitive to its electronic environment[263–267, 271–279]. Thus, ^{77}Se NMR spectroscopy offers the possibility to study synthetic selenium analogs of sulfur-containing natural products, enzymes which incorporate selenium as well as sulfur and naturally occurring selenoproteins.

Early continuous wave[280–282] and double resonance studies[271] examined a large number of small selenium-containing molecules with regard primarily to chemical shifts. These studies in general established the chemical shift ranges for various classes of organoselenium compounds. The most shielded selenium resonances are those attributed to anionic selenium compounds (e.g. $(NH_4)_2Se$, $CH_3Se^-Na^+$, etc.) and the most deshielded resonances are those from selenium-halogen and selenium-oxygen compounds and, interestingly, from selenium atoms which are doubly bonded to a carbon atom[278]. Although several reference compounds have been used by different groups, dimethyl selenide now appears to be the reference choice and the following general ranges of chemical shifts are found with respects to $(CH_3)_2Se$: selenols, +280 to −116 ppm; selenides, +425 to 0 ppm; and diselenides, +470 to +275 ppm.

Since nuclear spin-lattice relaxation times, T_1, are such a critical parameter in determining the recycle time of FT NMR experiments[283] several studies have examined the magnitude and mechanism of ^{77}Se relaxation[272, 273, 275, 284, 285]. For small molecules the spin rotation mechanism dominates and for larger molecules where this mechanism is not so effective, the chemical shift anisotropy mechanism becomes more effective. Interestingly, the dipole-dipole mechanism has not been found to be an efficient relaxation mechanism for any compound studied, even for selenols which contain a directly bound hydrogen.

With regard to biological systems, ^{77}Se NMR spectroscopy is still in its infancy. The amino acids selenomethione, selenocysteine and selenocystine have been studied as to chemical shift and/or relaxation times[273, 275)]. Recently the preparation of 6,6′-diselenobis-(3-nitrobenzoic acid) a selenium analogue of the well known Ellman's reagent was reported as were its properties for the quantitative estimation of sulfhydryl groups in proteins[286)]. It was demonstrated that the diselenide reacts specifically and quantitatively with thiol groups of proteins to yield a selenenyl sulfide. Subsequently, ribonuclease-A and lysozyme both of whose disulfide bonds had been reductively cleaved under denaturing conditions, were treated with this selenium reagent to yield covalently attached selenium which was then observed by high resolution ^{77}Se NMR spectroscopy[287)]. This was the first observation of ^{77}Se resonances in high molecular weight systems and demonstrated that ^{77}Se NMR signals emanating from ^{77}Se-containing biomolecules can be readily observed. This technique should be highly useful in studies of selenobiomolecules. Finally, in connection with the above study, it was necessary to define the ^{77}Se chemical shift range of organoselenenyl sulfides and a recent ^{77}Se NMR study of twenty-three of these compounds has appeared[288)].

6 Conclusions and Suggestions for Further Work

From the preceeding discussion of selenium biochemistry, it is obvious that our understanding of many aspects of this area exists at a very low level. Even though the toxicity of selenium and many of its compounds has been recognized for some time and may derive from selenium substitution for sulfur in many systems, the mechanism by which these adverse effects occur is still unknown. Further, naturally occurring selenoproteins are being discovered at a very rapid rate and the mechanism of the incorporation of selenium as well as the role of selenium in these systems is not well understood. Future research needs to focus on these mechanisms, on the roles and functions of selenium in biological processes and on a greater utilization of physical and spectroscopic techniques in these studies. Selenium-77 NMR spectroscopy should be especially well suited for these studies.

Acknowledgement. The author is especially grateful to Drs. Bruce Dunlap and Narender Luthra for many stimulating and helpful discussions concerning selenium biochemistry and to the National Institutes of Health for supporting biochemical selenium research at the University of South Carolina.

7 References

1. Cotton, F. A., Wilkinson, G.: Advanced Inorganic Chemistry, Wiley, New York, 1980, Chap. 16
2. Bagnall, K. W.: The Chemistry of Selenium, Tellurium and Polonium, Elsevier, Amsterdam, 1966
3. Kudryavtsev, A. A.: The Chemistry and Technology of Selenium and Tellurium (translated and revised by Elkin, E. M.), Collet's, London, 1974
4. Pauling, L.: The Nature of the Chemical Bond, Cornell University Press, Ithaca, N. Y., 1960, Chaps. 3 and 7
5. Bisbjerg, B.: Studies on Selenium in Plants and Soils, Danish Atomic Energy Commission, 1972
6. Rosenfeld, I., Beath, O. A.: Selenium Geobotany, Biochemistry, Toxicity and Nutrition, Academic Press, New York, 1964
7. Committee on Medical and Biologic Effects of Environmental Pollutants: Selenium, Nat. Acad. Sci., Washington, D. C., 1976
8. Sindeeva, N. D.: Mineralogy and Types of Deposits of Selenium and Tellurium, Wiley, New York, 1964
9. Greenland, L.: Geochim. Cosmochim. Acta *31,* 849 (1967)
10. Dufresne, A.: ibid. *20,* 141 (1960)
11. Schindewolf, U.: ibid. *19,* 134 (1960)
12. Suzuoki, T.: Bull. Chem. Soc. Jpn. *38,* 1940 (1965)
13. Suzuoki, T.: ibid. *38,* 1946 (1965)
14. Lakin, H. W., Davidson, D. F.: Symposium: Selenium in Biomedicine, (ed.) Muth, O. H., Avi Publishing Company, Westport, Connecticut, 1967, Chap. 3
15. Beath, O. A., et al.: J. Am. Pharm. Assoc. Sci. Ed. *23,* 94 (1934)
16. Moxon, A. L., Olson, D. E., Searight, W. V.: South Dakota Agr. Expt. Stn. Revised Techn. Bull. No. 2, 1 (1950)
17. Johnson, C. M., Asher, C. J. and Broyer, T. C.: Ref. 14, Chap. 4
18. Pillay, K. K. S., Thomas, C. C., Kamiaski, J. W.: Nucl. Applns. and Tech. *1,* 478 (1969)
19. Pidzhyan, G. O.: Izv. Akad. Nauk. Am. SSR, Nauki Zemle *20,* 81 (1967)
20. Davis, S. N., DeWiest, R. J. M.: Hydrogeology, Wiley, New York, 1966
21. Hashimoto, Y., Winchester, J. W.: Environ. Sci. and Tech. *1,* 338 (1967)
22. Ageton, R. W.: Mineral Facts and Problems. U.S. Bureau of Mines, U.S. Department of the Interior, Washington, D.C., Bulletin No. 650, 1970
23. Lo, M.-T., Sandi, E.: J. Environ. Path. Toxicol. *4,* 193 (1980)
24. Zabel, N. L. et al.: Am. J. Clin. Nutr. *31,* 850 (1978)
25. Schroeder, H. A., Mitchener, M.: J. Nutr. *101,* 1531 (1971)
26. Fishbein, L.: Toxicology of Selenium and Tellurium, in: Toxicology of Trace Elements (ed.) Goger, R. A. and Mehlman, M. A., p 191, Hemisphere Publ. Comp., Washington, 1977
27. Thomson, C. D., Burton, C. E., Robinson, M. F.: Brit. J. Nutr. *39,* 579 (1978)
28. Robinson, M. F. et al.: ibid. *39,* 589 (1978)
29. Nakamuro, K. et al.: Eisei Kagaku *20,* 75 (1974)
30. Lucky, T. D., Venugopal, B.: Metal Toxicity in Mammals, Plenum Press, New York, 1977
31. Glover, J. R.: Industr. Med. Surg. *39,* 50 (1970)
32. Broghamer, W. L., McConnell, K. P., Blatcky, A. L.: Cancer *37,* 1384 (1976)
33. Shamburger, R. J. et al.: J. Natl. Cancer Inst. *50,* 863 (1973)
34. Shamberger, R. J., Tytko, S., Willis, C. E.: Clin. Chem. *19,* 672 (1973)
35. Shamberger, R. J., Frost, D. V.: Can. Med. Assoc. *100,* 682 (1969)
36. Shamberger, R. J.: Proc. Am. Assoc. Cancer Res. *10,* 79 (1969)
37. Schrauzer, G. N.: Bioinorg. Chem. *5,* 275 (1976)
38. Schrauzer, G. N., White, D. A., Schneider, C. J.: ibid. *7,* 23 (1977)
39. Schrauzer, G. N., White, D. A., Schneider, C. J.: ibid. *7,* 35 (1977)
40. Schrauzer, G. N., Ishmael, D.: Ann. Clin. Lab. Sci. *4,* 441 (1974)
41. Selenium in Nutrition, Nat. Res. Council, Nat. Acad. Sci., Washington, D. C., 1971
42. Schwartz, K., Foltz, G. N.: J. Am. Chem. Soc. *79,* 3292 (1957)

43. Pedersen, N. D. et al.: Bioinorg. Chem. *2,* 33 (1972)
44. Klayman, D. L., Günther, W. H. H. (eds.): Organic Selenium Compounds: Their Chemistry and Biology, Wiley, New York, 1973
45. Zingaro, R. A., Cooper, W. C. (eds.): Selenium, Van Nostrand-Reinhold, New York, 1974, Chap. 8
46. Clive, D. L. J.: Modern Organoselenium Chemistry, Pergamon, Oxford, 1969
47. Reich, H. J. in: Oxidation in Organic Chemistry, (ed.) Trahanovsky, W. S., Academic Press, New York, 1978
48. Sharpless, K. B. et al.: Chem. Scr. *8A,* 9 (1975)
49. Clive, D. L. J.: Tetrahedron *34,* 1049 (1978)
50. Clive, D. L. J.: Aldrichimica Acta *11,* 43 (1978)
51. Reich, H. J.: Acc. Chem. Res. *12,* 22 (1979)
52. Ref. 44, Chap. XI
53. Chan, T. H., Finkenbine, J. R.: Tetrahedron Lett. 2091 (1974)
54. Clive, D. L. J., Denyer, C. V.: J. Chem. Soc., Chem. Commun., 253 (1973)
55. Van Ende, D., Krief, A.: Tetrahedron Lett., 2709 (1975)
56. Sharpless, K. B., Lauer, R. F.: J. Org. Chem. *37,* 3973 (1972)
57. Sharpless, K. B., Lauer, R. F.: J. Am. Chem. Soc. *94,* 7154 (1972)
58. Reich, H. J.: J. Org. Chem. *40,* 2570 (1975)
59. Reich, H. J., Shab, S. K.: J. Am. Chem. Soc. *99,* 263 (1977)
60. Reich, H. J. et al: J. Org. Chem. *43,* 1697 (1978)
61. Jones, D. N., Mundy, D., Whitehouse, R. D.: Chem. Commun., 86 (1970)
62. Gilman, H., Webb, F. J.: J. Am. Chem. Soc. *71,* 4062 (1949)
63. Gröbel, B.-T., Seebach, D.: Chem. Ber. *110,* 867 (1977)
64. Reich, H. J., Shal, S. K.: J. Am. Chem. Soc. *97,* 3250 (1975)
65. Mitchell, R. H.: J. Chem. Soc., Chem. Commun. 990 (1975)
66. Seebach, D., Peleties, N.: Chem. Ber. *105,* 511 (1972)
67. Seebach, D., Beck, A. K.: Angew. Chem., Int. Ed. Engl. *13,* 806 (1974)
68. Seebach, D., Meyer, N., Bech, A. K.: Liebigs Ann. Chem. *846*, 2217 (1977)
69. Reich, H. J., Ringa, J. M., Reich, I. L.: J. Am. Chem. Soc. *97,* 5434 (1975)
70. Barth, H. and Gosselck, J.: Z. Naturforsch. *166,* 280 (1961)
71. Pearson, R. G., Sobel, H., Songstad, J.: J. Am. Chem. Soc. *90,* 319 (1968)
72. Nakamura, N., Sekido, E.: Talanta *17,* 515 (1970)
73. Nylen, P.: Z. Anorg. Allg. Chem. *246,* 227 (1941)
74. Oki, M., Iwamura, H.: Tetrahedron Lett. 2917 (1966)
75. Stadtman, T. C.: Trends Biochem. Sci. *5,* 203 (1980)
76. Reference 45, Chap. 9
77. Challenger, F.: Adv. Enzymol. *12,* 429 (1951)
78. Ganther, H. E.: Biochemistry *5,* 1089 (1966)
79. Klug, H. L., Froom, J. D.: S. Dakota Acad. Sci. Proc. *64,* 247 (1965)
80. Evans, C. S., Asher, C. J., Johnson, C. M.: Austral. J. Biol. Sci. *21,* 13 (1968)
81. Byard, J. L.: Arch. Biochem. Biophys. *130,* 556 (1969)
82. Palmer, I. S. et al.: Biochim. Biophys. Acat *177,* 336 (1969)
83. McCready, R. G. L., Campbell, J. N., Payne, J. I.: Canad. J. Microbiol. *12,* 703 (1966)
84. Peterson, P. J., Butler, G. W.: Austral. J. Biol. Sci. *15,* 126 (1962)
85. Fredga, A.: Svensk Kem. Tidskr. *48,* 160 (1936)
86. Fredga, A.: ibid. *49,* 124 (1937)
87. Painter, E. P.: J. Am. Chem. Soc. *69,* 229 (1947)
88. Williams, L. R., Ravve, A.: ibid. *70,* 1244 (1948)
89. Janicki, J., Skopin, J., Zagalak, B.: Rocz. Chem. *36,* 353 (1962)
90. Frank, W.: Z. Physiol. Chem. *339,* 202 (1964)
91. Zdansky, G.: Ark. Kemi *17,* 273 (1961)
92. Chen, S.-H., Gunther, W. H. H., Mautner, H. G. in: Biochemical Preparations, (ed.) Brown, G. B., Wiley, New York, 1963, Vol. 10
93. Theodoropoulos, D., Schwartz, I. L., Walter, R.: Tetrahedron Lett. *25,* 2411 (1967)
94. Theodoropoulos, D., Schwartz, I. L., Walter, R.: Biochemistry *6,* 3927 (1967)
95. Roy, J., Gordon, W., Walter, R.: J. Org. Chem. *35,* 510 (1970)

96. Huber, R. E., Criddle, R. S.: Arch. Biochem. Biophys. *122,* 164 (1967)
97. Ref. 44, Chap XIIA
98. Oughton, B. M., Harrison, P. M.: Acta Cryst. *12,* 396 (1959)
99. Walter, R., Schlesinger, D. H., Schwartz, I. L.: Analyt. Biochem. *27,* 231 (1969)
100. Caldwell, K. A., Tappel, A. L.: Biochemistry *3,* 1643 (1964)
101. Nygard, B.: Ark. Kemi *27,* 341 (1967)
102. Zdansky, G.: ibid. *29,* 443 (1968)
103. Spare, C. G., Virtanen, A. I.: Acta Chem. Scand. *18,* 280 (1964)
104. Walter, R. in: Peptides: Chemistry and Biochemistry, (ed.) Weinstein, B., Marcel Dekker, New York, 1970
105. Trelease, S. F., Disomma, A. A., Jacobs, A. L.: Science *132,* 618 (1960)
106. Schrift, A., Varupaksha, T. K.: Biochem. Biophys. Acta *71,* 483 (1963)
107. Schrift, A., Varupaksha, T. K.: ibid. *100,* 65 (1965)
108. Stadtman, T. C.: Ann. Rev. Biochem. *49,* 93 (1980)
109. Horn, M. J., Jones, D. B.: J. Am. Chem. Soc. *62,* 234 (1940)
110. Virupaksha, T. K., Schrift, A.: Biochim. Biophys. Acta *74,* 791 (1963)
111. Peterson, P. J., Butler, G. W.: Nature *213,* 599 (1967)
112. Aronow, L., Kerdel-Vegas, F.: ibid. *205,* 1185 (1965)
113. Kerdel-Vegas, F. et al.: ibid. *205,* 1186 (1965)
114. Zdansky, G.: Ark. Kemi *29,* 449 (1968)
115. Pande, C. S., Rudick, J., Walter, R.: J. Org. Chem. *35,* 1440 (1970)
116. Painter, E. P.: J. Am. Chem. Soc. *69,* 232 (1947)
117. Klosterman, H. J., Painter, E. P.: ibid. *69,* 2009 (1947)
118. Zdansky, G.: Ark. Kemi *19,* 559 (1962)
119. Zdansky, G.: ibid. *29,* 437 (1968)
120. Jakubke, H.-D. et al.: Collect. Czeck. Chem. Commun. *33,* 3910 (1968)
121. Zdansky, G.: Ark. Kemi *21,* 211 (1963)
122. Virupaksha, T. K., Shrift, A., Tarver, H.: Biochim. Biophys. Acta *130,* 45 (1966)
123. Plieninger, H.: Chem. Ber. *83,* 265 (1950)
124. Zdansky, G.: Ark. Kemi *19,* 559 (1962)
125. Pan, F., Tarver, H.: Arch. Biochem. Biophys. *119,* 429 (1967)
126. Pan, F., Natori, Y., Tarver, H.: Biochim. Biophys. Acta *93,* 521 (1964)
127. Bremer, J., Natori, Y.: ibid. *44,* 367 (1960)
128. McConnell, K. P., Wabnitz, C. H.: ibid. *86,* 182 (1964)
129. Shepherd, L., Huber, R. E.: Can. J. Biochem. *47,* 877 (1969)
130. Virupaksha, T. K., Shrift, A.: Biochim. Biophys. Acta *107,* 69 (1965)
131. Stadtman, T. C.: Science *183,* 915 (1974)
132. Stadtman, T. C.: Adv. Enzymol. *48,* 1 (1979)
133. Schwarz, K., Folz, C. M.: J. Am. Chem. Soc. *79,* 3292 (1957)
134. Mills, G. C.: J. Biol. Chem. *229,* 189 (1957)
135. Rotruck, J. T. et al.: Fed. Proc. *31,* 691 (abstr.) (1972)
136. Rotruck, J. T. et al.: Science *179,* 588 (1973)
137. Lehninger, A. L.: Biochemistry, Worth Publishers, New York, 1975[2]
138. Lynen, F., Reichert, E. and Rueff, L.: Justus Liebigs Ann. Chem. *574,* 1 (1951)
139. Lynen, F., Reichert, E.: Angew. Chem. *63,* 47 (1951)
140. Gunther, W. H. H., Mautner, H. G.: J. Am. Chem. Soc. *87,* 2708 (1965)
141. Knappe, J.: Ann. Rev. Biochem. *39,* 757 (1970)
142. Moss, J., Lane, D.: Adv. Enzymol. *35,* 321 (1971)
143. Alberts, A. W., Vagelos, P. R. in: The Enzymes, (ed.) Boyer, P. D., Academic Press, New York, 1972[3], p. 37
144. Marx, M. et al.: J. Am. Chem. Soc. *99,* 6754 (1977)
145. Confalone, P. N. et al.: ibid. *99,* 7020 (1977)
146. Bory, S., Marquet, A.: Tetrahedron Lett. 2033 (1976)
147. Piffeateau, A., Gaudry, M., Marquet, A.: Biochem. Biophys. Res. Commun. *73,* 773 (1976)
148. Marquet, A.: Pure Appl. Chem. *49,* 183 (1977)
149. Lindblow-Kull, C., Kull, F., Shrift, A.: Biochem. Biophys. Res. Commun. *93,* 572 (1980)
150. Orme-Johnson, W. H. et al.: Proc. Natl. Acad. Sci. U.S. *60,* 368 (1968)

151. Fee, J. A., Palmer, G.: Biochem. Biophys. Acta *245,* 175 1971)
152. Mukai, K. Huang, J. J., Kimura, T.: Biochem. Biophys. Acta *336,* 427 (1974)
153. Pal, B. C., Schmidt, D. G.: J. Am. Chem. Soc. *99,* 1973 (1977)
154. Christ, W., Rakow, D., Coper, H.: Fresenius Z. Anal. Chem. *279,* 159 (1976)
155. Busiello, V. et al.: Biochim. Biophys. Acta *606,* 347 (1980)
156. DeMarco, C. et al.: ibid. *478,* 156 (1977)
157. Aherne, W., Piall, E., Marks, V.: Ann. Clin. Biochem. *15,* 331 (1978)
158. Ref. 44, Chap. XIIIE
159. Huber,R. E. Criddle, R. S.: Biochim. Biophys. Acta *141,* 587 (1967)
160. Coch, E. H., Greene, R. C.: ibid. *230,* 223 (1971)
161. Young, P. A., Kaiser, I. I.: Arch. Biochem. Biophys. *171,* 483 (1975)
162. Shrift, A. et al.: Plant Physiol. *58,* 248 (1976)
163. Lipsett, M. N.: J. Biol. Chem. *240,* 3975 (1965)
164. Abrell, J. W., Kaufman, E. F., Lipsett, M. N.: ibid. *246,* 294 (1971)
165. Dilworth, G. L., Bandurski, R. S.: Biochem. J. *163,* 521 (1977)
166. Mudd, S. H., Cantoni, G. L.: Nature *180,* 1052 (1957)
167. Turner, D. C., Stadtman, T. C.: Arch. Biochem. Biophys. *154,* 366 (1973)
168. Cone, J. E., Martin del Rio, R., Stadtman, T. C.: J. Biol. Chem. *252,* 5337 (1977)
169. Beilstein, M. A., Tripp, M. J., Whanger, P. D.: J. Inorg. Biochem. *15,* 339 (1981)
170. Junde, R. A., Hockstra, W. G.: Biochem. Biophys. Res. Commun. *93,* 1181 (1980)
171. Hawkes, W. C., Lyons, D. E., Tappel, A. L.: Fed. Proc. *38,* 820 (abstr.) (1979)
172. Yasumoto, K., Kimikazu, I., Yoshida, M.: J. Nutr. *109,* 760 (1979)
173. Flohe, L., Gunzler, W. A., Schock, H.: FEBS Lett. *32,* 132 (173)
174. Flohe, L. et al.: Hoppe-Seylers Z. physiol. Chem. *353,* 987 (1972)
175. Schneider, F., Flohe, L.: ibid. *348,* 540 (1967)
176. Flohe, L., Gunzler, W. A. in: Glutathione, (ed.) Jakoby, W. B., Academic Press, New York, 1974
177. Mills, G. C., Randall, H. P.: J. Biol. Chem. *232,* 589 (1958)
178. O'Brien, P. J., Little, C.: Can. J. Biochem. *47,* 493 (1969)
179. Flohe, L., Eisele, B., Wendel, A.: Hoppe-Seylers Z. Physiol. Chem. *352,* 151 (1971)
180. Oh, S.-H., Ganther, H. E., Hoekstra, W. G.: Biochemistry *13,* 1825 (1974)
181. Awasthi, Y. C., Beutler, E., Srivastava, S. K.: J. Biol. Chem. *250,* 5144 (1975)
182. Nakamura, W., Hosoda, S., Hayashi, K.: Biochim. Biophys. Acta *358,* 251 (1974)
183. Forstrom, J. W., Zakowski, J. J., Tappel, A. L.: Biochemistry *17,* 2639 (1978)
184. Ladenstein, R. et al.: J. Mol. Biol. *134,* 199 (1979)
185. Holmberg, N. J.: Exp. Eye. Res. *7,* 570 (1968)
186. Pedersen, N. D. et al.: Bioinorg. Chem. *2,* 33 (1972)
187. Hartley, W. J., Grant, A. B.: Fed. Proc. *20,* 679 (1961)
188. Schubert, J. R. et al.: ibid. *20,* 689 (1961)
189. Black, R. S. et al.: Bioinorg. Chem. *8,* 161 (1978)
190. Beilstein, M. A., Tripp, M. J., Whanger, P. D.: Fed. Proc. *39,* 388 (abstr.) (1980)
191. McConnell, K. P. et al.: Biochim. Biophys. Acta *588,* 113 (1979)
192. Enoch, H. G., Lester, R. L.: J. Biol. Chem. *250,* 6693 (1975)
193. Andreesen, J. R., Ljungdahl, L. G.: J. Bacteriol. *116,* 817 (1973)
194. Wagner, R., Andreesen, J. R.: Arch. Microbiol. *114,* 219 (1977)
195. Andreesen, J. R., El Ghazzawi, E., Gottschalk, G.: ibid. *96,* 103 (1973)
196. Leonhardt, U., Andreesen, J. R.: ibid. *115,* 277 (1977)
197. Stadtman, T. C.: Nutr. Rev. *35,* 161 (1977)
198. Jones, J. B., Dilworth, G. L., Stadtman, T. C.: Arch. Biochem. Biophys. *195,* 255 (1979)
199. Kröger, A. et al.: Eur. J. Biochem. *94,* 467 (1979)
200. Ohyama, T., Yamazaki, I.: J. Biochem. Tokyo *75,* 1257 (1974)
201. Tanaka, H., Stadtman, T. C.: J. Biol. Chem. *254,* 447 (1979)
202. Stadtman, T. C.: Arch. Biochem. Biophys. *113,* 9 (1966)
203. Cone, J. E. et al.: Proc. Natl. Acad. Sci. U.S.A. *73,* 2659 (1976)
204. Harary, I.: J. Biol. Chem. *227,* 815 (1957)
205. Pastan, I., Tsai, L., Stadtman, E. R.: ibid. *239,* 902 (1964)
206. Harary, I.: ibid. *227,* 823 (1957)

207. Holcenberg, J. S., Stadtman, E. R.: ibid. *244,* 1194 (1969)
208. Imhoff, D., Andreesen, J. R.: FEMS Microbiol. Lett. *5,* 155 (1979)
209. Hartmanis, M.: Abstr. 64th Meet. Amer. Soc. Biological Chem., New Orleans, LA, 1980
210. Goldman, P., Alberts, A. W., Vagelos, P. R.: Biochem. Biophys. Res. Commun. *5,* 280 (1961)
211. Wagner, R., Andreesen, J. R.: Arch. Microbiol. *121,* 255 (1979)
212. Bradshaw, W. H., Barker, H. A.: J. Biol. Chem. *235,* 3620 (1960)
213. Reference 44, Chap. XV A
214. Sharghi, N., Lalezari, I.: Spectrochim Acta *20,* 237 (1964)
215. Harvey, A. B., Wilson, M. K.: Inorg. Nucl. Chem. Lett. *1,* 101 (1965)
216. Harvey, A. B., Wilson, M. K.: J. Chem. Phys. *45,* 678 (1966)
217. Merijanian, A. Zingaro, R. A., Sagan, L. S., Irgolic, K. J.: Spectrochim. Acta *25 A,* 1160 (1966)
218. Allkins, J. R., Hendra, P. J.: ibid. *22,* 2075 (1966)
219. Greenwood, N. N., Hunter, G.: J. Chem. Soc., A. 929 (1969)
220. Bahr, K. W., Fowles, G. W. A.: ibid. 801 (1968)
221. Hendra, P. J., Sadasivan: Spectrochim. Acta *21,* 1127 (1965)
222. Aynsley, E. E.: Chem. Ind. (London), 379 (1966)
223. Hamada, K., Morishita, H.: Spectroscopy Lett. *10,* 367 (1977)
224. Paetzold, R. et al.: Z. Anorg. Allg. Chem. *352,* 295 (1967)
225. Lopez, J., Jao, T. C., Rudzinski, W. E.: J. Inorg. Biochem. *14,* 177 (1981)
226. Green, W. H., Harvey, A. B.: J. Chem. Phys. *49,* 3586 (1968)
227. Allum, K. G. et al.: Spectrochim. Acta *24 A,* 927 (1968)
228. Bergson, G.: Ark. Kemi *13,* 11 (1959)
229. Günther, W. H. H.: J. Org. Chem. *32,* 3931 (1967)
230. Noda, L. H., Kudy, S. A., Lardy, H. A.: J. Am. Chem. Soc. *75,* 913 (1953)
231. Chierici, L., Passerini, R.: Atti Accad. Naz. Lincei, Rend., Classe Sci. Fis. Mat. Nat. *14,* 99 (1953)
232. Kiss, A. I., Muth: Acta Chim. Acad. Sci. Hung. *22,* 396 (1960)
233. Bogolyubov, G. M., Shlyk, Y. N.: Zh. Obshch. Khim. *39,* 1759 (1969)
234. Ref. 44, Chap. XV B
235. Fredga, A.: Svensk Kem. Tidskr. *48,* 91 (1936)
236. Bergson, G.: Ark. Kemi *13,* 11 (1958)
237. Bergson, G., Claeson, G., Schotte, L.: Acta Chem. Scand. *16,* 1159 (1962)
238. Linderberg, J., Michl, J.: J. Am. Chem. Soc. *92,* 2619 (1970)
239. Pauling, L.: Proc. Nat. Acad. Sci. U.S. *35,* 495 (1949)
240. Bergson, G.: Ark. Kemi *19,* 195 (1962)
241. Bergson, G.: ibid. *9,* 121 (1955)
242. Bergson, G., Wold, S.: ibid. *19,* 215 (1962)
243. Ref. 44, Chap. XV C
244. Djerassi, C., Fredga, A., Sjoberg, B.: Acta Chem. Scand. *15,* 417 (1961)
245. Djerassi, C., Wolf, H., Bunnenberg, E.: J. Am. Chem. Soc. *84,* 4552 (1962)
246. Walter, R., du Vigneaud, V.: ibid. *87,* 4192 (1965)
247. Walter, R., du Vigneaud, V.: ibid. *88,* 1331 (1966)
248. Walter, R., Chan, W. Y.: ibid. *89,* 3892 (1967)
249. Urry, D. W. et al.: Proc. Natl. Acad. Sci. U.S. *60,* 967 (1968)
250. Craig, J. C. et al.: J. Am. Chem. Soc. *98,* 6456 (1976)
251. Blaha, K., Fric, I., Jakubke, H.-D.: Collect. Czech. Chem. Commun. *32,* 558 (1967)
252. Ref. 44, Chap. XV H
253. Shefter, E., James, M. N. G., Mautner, H. G.: J. Pharm. Sci. *55,* 643 (1966)
254. George, C. et al.: J. Chem. Phys. *55,* 1071 (1971)
255. Palmer, H. T., Palmer, R. A.: Acta Cryst. *25 B,* 1090 (1969)
256. Marsh, R. E.: ibid. *5,* 458 (1952)
257. Kruse, F. H., Marsh, R. E., McCullough: ibid. *10,* 201 (1957)
258. Valle, G. et al.: ibid. *B 26,* 468 (1970)
259. Rutherford, J. S. and Calvo, C.: Z. Krist. *128,* 229 (1969)
260. Hope, H.: Acta Cryst. *18,* 259 (1965)

261. Ref. 44, Chap. XVD
262. Rodger, C. et al. in: NMR and the Periodic Table, Academic Press, New York, 1978
263. Fredga, A., Gronowitz, S., Hornfeldt, A.-B.: Chem. Ser. *8,* 15 (1975)
264. Gronowitz, S., Johnson, I., Hornfeldt, A.-B.: ibid. *3,* 94 (1973)
265. Gronowitz, S., Johnson, I., Hornfeldt, A.-B.: ibid. *8,* 8 (1975)
266. Gronowitz, S., Komar, A., Hornfeldt, A.-B.: Org. Magn. Reson. *9,* 213 (1977)
267. Christiaens, L. et al.: Org. Magn. Reson. *8,* 354 (1976)
268. Shulman, R. G. (ed.): Biological Applications of Magnetic Resonance, Academic Press, New York, 1979
269. Dwek, R. A.: Nuclear Magnetic Resonance in Biochemistry Applications to Enzyme Systems, Oxford Univ. Press, London, 1973
270. Wuthrich, K.: NMR in Biological Research: Peptides and Proteins, North Holland Press, Amsterdam, 1976
271. McFarlane, W. M., Wood, R. J.: J. Chem. Soc., A. 1397 (1972)
272. Dawson, W. H., Odom, J. D.: J. Am. Chem. Soc. *99,* 8352 (1977)
273. Odom, J. D., Dawson, W. H., Ellis, P. D.: ibid. *101,* 5815 (1979)
274. Jakobsen, H. J. et al.: J. Magn. Reson. *38,* 219 (1980)
275. Pan, W.-H., Fackler, J. D.: J. Am. Chem. Soc. *100,* 5383 (1978)
276. Kalabin, G. A. et al.: Org. Magn. Reson. *12,* 598 (1979)
277. Granger, O., Chapelle, S., Paulmier, C.: Org. Magn. Reson. *14,* 240 (1980)
278. Cullen, E. R. et al.: J. Am. Chem. Soc. *103,* 7055 (1981)
279. Llabres, G. et al.: Org. Magn. Reson. *16,* 17 (1981)
280. Birchall, T., Gillespie, R. J., Vekris, S. L.: Can. J. Chem. *43,* 1672 (1965)
281. Lardon, M.: J. Am. Chem. Soc. *92,* 1063 (1970)
282. Lardon, M.: Ann. N. Y. Acad. Sci. *192,* 132 (1972)
283. Farrar, T. C., Becker, E. D.: Pulse and Fourier Transform NMR, Academic Press, New York, 1971
284. Gansow, O. A., Vernon, W. D., Dechter, J. J.: J. Magn. Reson. *32,* 19 (1978)
285. Koch, W., Lutz, O., Nolle, A.: Z. Naturforsch. *33A,* 1025 (1978)
286. Luthra, N. P., Dunlap, R. B., Odom, J. D.: Anal. Biochem. *117,* 94 (1981)
287. Luthra, N. P. et al.: J. Biol. Chem. *257,* 1142 (1982)
288. Luthra, N. P., Dunlap, R. B., Odom, J. D.: J. Magn. Reson. *46,* 152 (1982)

The Ribonucleotide Reductases – A Unique Group of Metalloenzymes Essential for Cell Proliferation

Dedicated to Professor Peter Karlson on the occasion of his 65. birthday

Manfred Lammers and Hartmut Follmann

Fachbereich Chemie, Arbeitsgruppe Biochemie der Philipps-Universität, D-3550 Marburg, Fed. Rep. Germany

Reductive elimination of the 2′-hydroxyl group from ribonucleotides to yield 2′-deoxyribonucleotides, the monomeric precursors of DNA, requires an uncommon type of enzyme catalysis in which the transition metals, manganese, iron, or cobalt, and free radical intermediates cooperate. In the group of deoxyadenosylcobalamin (coenzyme B 12)-dependent ribonucleotide reductases the coenzyme supplies a transient radical pair of deoxyadenosyl· and cob(II)alamin whereas in the non-heme-iron group of enzymes a protein subunit carries a stable tyrosyl radical coordinated to a binuclear iron(III) complex; the manganese-dependent enzymes are less precisely known. The radicals are thought to function in hydrogen transfer from cysteine SH to the ribonucleotide substrate, and the metal complexes are apparently needed to generate and stabilize the radicals. In aerobic organisms oxygen also plays a critical role in these processes and hence in DNA synthesis and cell proliferation. In addition to the transition metals, Mg^{2+} or Ca^{2+} are required by several ribonucleotide reductases for structural integrity. However the most potent inhibitors of deoxyribonucleotide biosynthesis (of potential interest in chemotherapy) are not metal chelators but radical scavengers. Cell cycle arrest and cell death produced by simple chemicals like N-hydroxyurea and hydroxamates can be traced back to their reaction with ribonucleotide reductase. Evidence is accumulating that independent enzymes of deoxyribonucleotide and DNA synthesis are functionally coupled in a novel type of supramolecular structure.

Structure and Bonding 54

A. Introduction

If one argues that deoxyribonucleic acid is the molecular master key to the existence and propagation of all living things in general and of each individual organism, one has to add that deoxyribonucleotides are just as essential for life. These small, unpretentious DNA precursor molecules are not found in more than trace amounts anywhere in nature; their chemistry and biochemistry, however, has long time not by far been as popular as that of the famous DNA macromolecule and its replication machinery. Discovered in the early 1950s and described in many details thence by Peter Reichard[1–4] the pathway of reductive formation of 2′-deoxyribonucleotides from ribonucleotides was only in the last decade recognized as a checkpoint for cell proliferation.

ⓅⓅOCH_2 O base, OH OH + SH SH → ⓅⓅOCH_2 O base, OH H + S–S + H_2O (I)

ribonucleotide + dithiol (e.g., reduced thioredoxin) → 2′-deoxyribonucleotide + disulfide (e.g., oxidized thioredoxin) + H_2O

Two aspects provoke special interest in the process expressed in equation I:
- Although the redox reaction (I) looks chemically simple, not requiring intramolecular rearrangements or an extra energy source, it has been and still is hard enough to explore its mechanism. No reasonable model reaction has yet become known, nor is enzyme catalysis fully understood. The small group of proteins catalyzing that irreversible hydrogen transfer, ribonucleotide reductases (EC 1.17.4), form an unusual, probably unique set of enzymes with structurally different yet functionally related metal-containing catalytic centers. Such properties not only pertain to present-day biochemistry but also raise questions about the prebiotic emergence of DNA.
- Although deoxyribonucleoside 5′-phosphates serve "only" as DNA building blocks in cellular metabolism and rarely accumulate, they nevertheless exert other, physiologically important as well as unprogrammed actions during biosynthesis and while they exist in minute, transient concentration. Deoxyribonucleoside triphosphates (dNTP) allosterically affect their own formation, thus controlling the rate and extent of DNA synthesis from the beginning. Disturbances of the delicate balance between all four DNA precursors can have serious consequences: dATP and dGTP are, for example, toxic metabolites in the molecular pathogenesis of human immunodeficiency diseases[5], and they appear to be significant targets for chemical mutagenesis induced by N-methyl-N-nitroso urea[6]. Since depletion of DNA precursors halts cell proliferation, inhibitors of ribonucleotide reduction are under study as potential drugs in antineoplastic and antiviral chemotherapy.

Due to the central importance and the ramifications of deoxyribonucleotide formation far beyond reaction (I) the literature dealing with ribonucleotide reduction is now rapidly expanding. Different aspects, in particular the complex allosteric and cellular regulation phenomena, the involvement of vitamin B 12, and evolutionary questions have been reviewed in recent years[7–10, 145, 354]. In this article we concentrate on the available information about various enzyme systems, their mechanisms and inhibitors, in an attempt to critically evaluate some basic essentials which in turn are prerequisite for understanding many of the biological consequences; the latter are dealt with more briefly. Of course, as everybody engaged in the field will probably agree, ribonucleotide reductase research is so full of conceptual and experimental intricacies that our present approach may at best put into place another few pieces of a puzzle.

B. Nature and Distribution of Ribonucleotide-Reducing Enzymes

It should be realized at the outset that all organisms have to possess the capacity to make deoxyribonucleotides from ribonucleotides. This is the only process which permits the cell to utilize one fraction of the total nucleotides formed de novo in pyrimidine and purine biosynthesis for DNA replication; there is no alternative biochemical route producing 2-deoxyribose, its phosphates, or N-glycosides from other molecules (Scheme II).

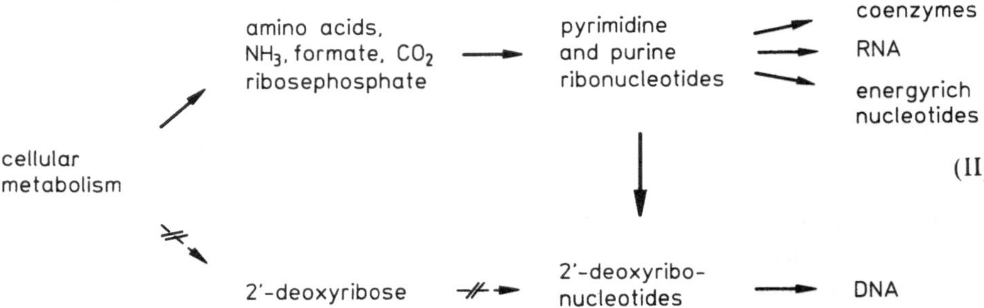

(II)

Such universal distribution and distinct function does not mean that the enzymes could be readily studied everywhere. Due to their cell cycle dependence described later and rather poor activity in cell-free extracts ribonucleotide reductases can only be purified and characterized under carefully chosen conditions. Therefore the organisms named as enzyme source in this chapter are not too numerous. They include some obvious, biochemically well-known candidates for the study of DNA precursors like *Escherichia coli*, yeast, or calf thymus, while other, more remote ones (e.g., *Brevibacterium ammoniagenes* or the alga, *Euglena gracilis*) have been added because of some peculiarity of their DNA synthesis. A systematic view of enzymatic deoxyribonucleotide formation is assembled in Table 4.

All ribonucleotide reductases are of complex protein structure and mechanism. At present we know three types which differ in their cofactor and metal requirement despite identical function. These are

1. bacterial and cyanobacterial enzymes depending on 5′-deoxyadenosylcobalamin, the cobalt-containing coenzyme form of vitamin B 12, for activity;
2. bacterial, viral, and eukaryotic metalloenzymes that contain protein-bound iron as part of the catalytic site; and
3. an enzyme group – of as yet uncertain distribution – where manganese is involved in ribonucleotide reduction.

Within these groups there may be further substantial structure differences among individual enzymes. The long-held view that *Lactobacillus leichmannii* ribonucleotide reductase is typical of all B 12-dependent, and *E. coli* reductase of all other enzymes including the eukaryotic ones appears no longer warranted.

1. Deoxyadenosylcobalamin Dependence

The fermentative bacteria, *Lactobacillus leichmannii* probably possess the most simple of all ribonucleotide reductases, but the protein does require the most complex coenzyme known, 5′-deoxyadenosylcobalamin (Fig. 1). This corrinoid coenzyme catalyzes but few enzymatic reactions, usually involving an intramolecular 1,2-hydrogen shift as found in the common methylmalonylCoA $\rightarrow$ succinylCoA isomerase reaction. Participation in deoxyribonucleotide formation, where hydrogen transfer occurs intermolecularly (Eq. I), was first implicated by the nutritional requirement of lactobacilli for vitamin B 12 during DNA synthesis, and was verified in cell-free extracts by Blakley and H. A. Barker in 1964[11)]. The reaction has been studied in great detail in the laboratories of Beck[13, 16)], Blakley[12, 14)], and Hogenkamp[9, 17, 153)].

L. leichmannii ribonucleotide reductase may only in comparison with the iron-enzymes be called a "simple" protein, but in fact is a polypeptide of unusual size. Homogenous enzyme has been prepared in 10 mg amounts, starting from kilogram quantities of bacteria, by classical procedures or by affinity chromatography on a Sepharose-immobilized effector nucleotide, dGTP[14, 15)]. It has a molecular weight of 76,000 and consists of approximately 690 amino acids in a single polypeptide chain. No subunits could be detected during sedimentation or by gel electrophoresis under denaturing condi-

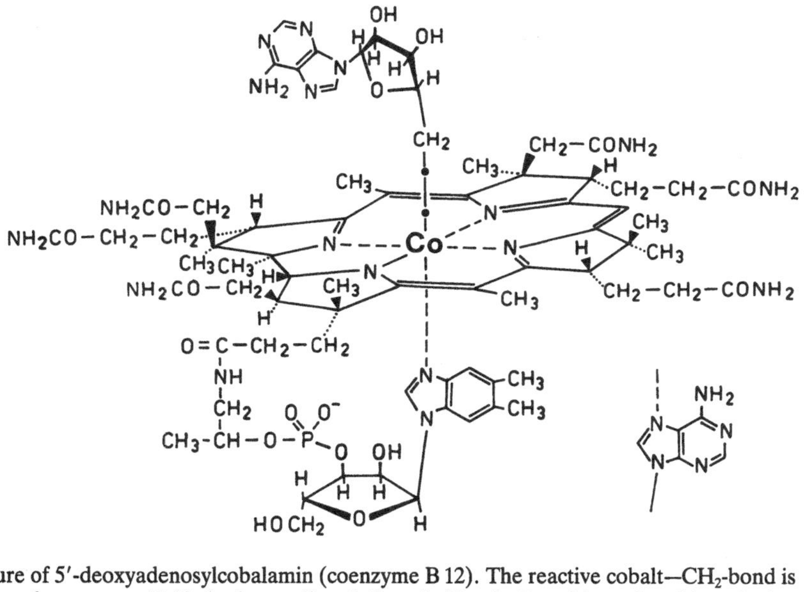

Fig. 1. Structure of 5′-deoxyadenosylcobalamin (coenzyme B 12). The reactive cobalt—CH_2-bond is marked. In pseudocoenzyme B 12 the lower dimethylbenzimidazole ligand is replaced by adenine (*lower right*)

tions, and the monomeric structure was also confirmed by establishing serine and lysine as the only N- and C-terminal amino acid, and by cyanogen bromide cleavage into the nine fragments expected from a content of eight methionines. The amino acid analysis[14)] shows no special features but sequencing of the large protein has not yet been undertaken.

Large size of the apoenzyme is understandable because it has to specifically interact with at least four other, different molecules, viz. the coenzyme as hydrogen carrier, a dithiol as hydrogen donor, a substrate ribonucleoside triphosphate, and one or two other nucleoside triphosphates serving as allosteric effectors. Structural requirements for mutual binding of all these components will be discussed in Sect. C. It may be noted here that apoenzyme-coenzyme interaction is particularly critical for catalysis and involves substituents at both axial cobalt ligands, dimethylbenzimidazol and 5′-deoxyadenosine, as well as the properly amidated side chains at the periphery of the corrin ring[7)] (Fig. 1). Moreover during catalysis a transient radical of "active coenzyme B 12" appears[161)]. In contrast to many other ribonucleotide reductases, divalent cations like Mg^{++} do not significantly affect structure and kinetics of the *Lactobacillus* enzyme system.

B 12-dependence of deoxyribonucleotide formation is fairly common among microorganisms. Knowledge of this group of ribonucleotide reductases has profited greatly from the simple, selective, and apparently generally applicable "tritium exchange reaction" in which a cell extract is incubated with a dithiol, an effector nucleotide of ribonucleotide reduction (usually ATP), and synthetic [5′-^{3}H]deoxyadenosylcobalamin. If a B 12-dependent reductase is present, interaction with the above components will result in the liberation of tritium from labeled coenzyme to water which can be readily determined[16, 17)]. The mechanism of this reaction is depicted in Fig. 12. If corrected for a low level of non-enzymatic tritium release by appropriate controls, enzyme activity can be estimated even in the absence of true substrate reduction and without further purification. Gleason has surveyed many organisms in this way[18–20)] and found no evidence for B 12-dependence in extracts from animal cells or typical algae but positive results in many strains of bacteria, cyanobacteria (the blue-green algae), and in the eukaryotic Euglenophyta (cf. Table 4). The B 12-dependent ribonucleotide reductases of *Euglena gracilis, Corynebacterium nephridii, Thermus aquaticus*, and *Anabaena sp.* have then been purified and are described in some detail below. Root nodule bacteria (rhizobia) also possess such an enzyme. The validity of the tritium exchange assay is supported by its failure in such instances where the iron- or manganese-type of reductase has been identified, e.g. in the green algae, in *E. coli*, and in certain *Micrococcus* and *Synechococcus* species.

Two other ribonucleotide reductases possess molecular weights and monomeric structure comparable to that of the *Lactobacillus* enzyme. One is the protein partially purified from an extreme thermophile, *Thermus aquaticus* (X−1) for which $M_r \approx 80{,}000$ was established by gel filtration[21)]. This ribonucleoside triphosphate reductase presents a very interesting example of thermostable *and* allosterically controlled enzyme. It shows maximum activity at 70 °C in vitro, close to the optimum growth temperature. Allosteric stimulation of ATP reduction by dGTP becomes effective only at temperatures higher than 55 °C. Thus if the mechanism of activation by effector nucleotides involves changes of protein conformation there must exist more than one energetically favoured active site conformation even at elevated temperature. Evidence for temperature and nucleotide-dependent conformational change was in fact deduced from the kinetic behaviour but the system has not been analyzed further. Mg^{++} or Mn^{++} ions, when present at equimolar concentration with substrate nucleotides stimulate enzyme activity but the reductase does not show an absolute metal requirement.

Metal ions are clearly essential for the ribonucleoside triphosphate reductase isolated from the filamentous cyanophyte, *Anabaena* 7119, one of the many blue-green algae that depend on deoxyadenosylcobalamin for deoxyribonucleotide synthesis[19, 22)]. The purified enzyme possesses a molecular weight of 72,000 (estimated by gel filtration) with no subunit structure. It does not reduce ribonucleotides in the absence of divalent cations; Ca^{++} is most effective but Mg^{++} and Mn^{++} also support enzyme catalysis. Judging from their optimum concentration (5–10 mM) the metal ions are not only necessary to complex the substrate triphosphate but should have an effect on the enzyme protein itself.

Next in complexity to the $M_r \approx 76{,}000$ ribonucleotide reductases is the enzyme purified to homogeneity from extracts of anaerobically grown *Corynebacterium nephridii*[23)]. It is a diphosphate reductase which is obtained by column chromatography as a protein of $M_r = 200{,}000$; under denaturing conditions, subunits of $M_r = 100{,}000$ are observed. As the N-terminus gave only one amino acid sequence it appears certain that *Corynebacterium* reductase is composed of two identical subunits. It does not require Mg^{++} and at >5 mM concentration the metal is inhibitory. The reaction is absolutely dependent on 5′-deoxyadenosylcobalamin but the number of coenzyme molecules per native holoenzyme (1, or 2?) is unknown; it may be difficult to determine in view of the nucleotide-dependence of coenzyme affinity. Another, probably oligomeric form of active enzyme tightly bound to a dGTP-Sepharose affinity column also awaits further characterization.

The eukaryotic algae, *Euglena gracilis*, systematically remote from other green algae, can serve as test organism in microbiological vitamin B 12 assays. Its cobalamin-dependent ribonucleotide reductase has been purified by several groups[24–26)]. These studies agree in the substrate triphosphate specificity and in that 2 mM Mg^{++} has slight, if any effect on the enzyme. However the question of size and subunit structure remains unsettled; Gleason and Hogenkamp found an active protein of $M_r = 145{,}000$ by sedimentation in a sucrose density gradient[24)] while Hamilton observed a large enzyme species of 440,000 dalton during gel filtration and a single protein migrating at $M_r = 100{,}000$ on SDS-polyacrylamide gels[25)]. Considering the not strictly comparable methods the data can probably be reconciled with the existence of a $\approx 100{,}000$ dalton subunit that dimerizes or tetramerizes in solutions of different composition. As B 12-deficient *Euglena* cells appear to accumulate the apoenzyme[26)], a source for more detailed structure analysis is available.

In a similar manner, cobalt deficiency was exploited for growth of reductase-overproducing *Rhizobium meliloti* cells[27, 28)] and subsequent purification of deoxyadenosylcobalamin-dependent ribonucleoside diphosphate reductase. The homogenous enzyme is a large protein but its molecular mass difficult to establish due to concentration-dependent aggregation and disaggregation effects. Sedimentation velocity experiments yield an extrapolated value of slightly over 500,000, and SDS polyacrylamide gel electrophoresis indicate the existence of 130,000 and 110,000 dalton proteins. Thus, the structure of *Rhizobium* ribonucleotide reductase is possibly also tetrameric. Divalent cations (Mg^{++}, Ca^{++}, Mn^{++}, Zn^{++} or Cd^{++}) inhibit the enzyme but in presence of an effector nucleotide like dGTP the inhibitory effects of Mg^{++} and Ca^{++} are relieved[28)]. Although these interactions indicate very complex kinetic behavior it may be stated that at least free metal ions are not essential for enzyme activity.

2. The Non-Heme-Iron Protein of *Escherichia coli*

After more than 20 years of intense research, predominantly carried out at the Karolinska Institute of Stockholm by Reichard, Thelander, Sjöberg and others, ribonucleoside diphosphate reductase of the well-known enterobacterium *E. coli* is certainly the best studied enzyme of its kind. In fact, it was in this organism that a complete set of four proteins capable to utilize NADPH for ribonucleotide reduction in vitro was first recognized, viz. two nonidentical ribonucleotide reductase subunits, thioredoxin, and thioredoxin reductase serving for hydrogen transfer[29]. The ribonucleotide reductase genes of *E. coli* (*nrdA, nrdB*) have now been integrated into a lysogenic phage[30] as well as into recombinant plasmids[31], and the enzyme protein has been investigated by any available physical method including Mößbauer spectroscopy; it may seem characteristic for the complexity of ribonucleotide reductases that despite such enormous progress the multiple interactions of protein subunits, metal ions, substrate and effector nucleotides are still not known to the atomic scale.

Pure ribonucleotide reductase is best prepared from a thymine-starved strain, *E. coli* B 3[32], or in even larger quantity from overproducing strain KK 546 (λ dnrd) lysogenic for a transducing phage that carries both structural genes attached to the λ genome. Approximately 15% of the total protein of such cells are ribonucleotide reductase, 50 times more than can be found in wild type *E. coli*[30]. Detection and purification is also aided by the availability of antibodies and chromatography on antibody (γ-globulin)-Sepharose columns.

A characteristic property of the *E. coli* enzyme is its easy dissociation into two protein fractions, B 1 and B 2, which are inactive separately but restore enzyme activity upon combination in the presence of Mg^{++}. The two non-identical subunits separate already during chromatography on DEAE cellulose[32] as well as during affinity chromatography on dATP-Sepharose[33] or similar materials; immobilized nucleotide ligands bind protein B 1 while B 2 is not retained on the column. Both were characterized in great detail[33] and were shown to possess a molecular weight of 160,000 (B 1) and 78,000 (B 2), respectively. In accord with the value expected for a 1 : 1 complex, sedimentation analysis of active holoenzyme in the presence of 15 mM Mg^{++} and 0.1 mM of the positive effector thymidine triphosphate indicated a single species with a sedimentation coefficient of 10 S and a molecular weight of 245,000. Heterogenous aggregates appeared in sucrose solutions and in presence of the negative effector, dATP. Each subunit is in turn composed of two polypeptide chains which separate only when denatured with guanidinium chloride or sodium dodecyl sulfate: B 1 then yields a molecular weight of 80,000 and B 2 one of 39,000. The two chains of B 1 are similar or identical in size but appear to differ somewhat in their N-terminal amino acid sequence. Thus, the overall composition of *E. coli* ribonucleotide reductase is of the type $\alpha\alpha'\beta_2$. A schematic representation of the holoenzyme is reproduced in Fig. 2. Unfortunately the enzyme subunits could not as yet be obtained in crystals of sufficient size or stability for X-ray structure analysis.

Subunit B 1 of *E. coli* ribonucleotide reductase is an SH protein carrying several binding sites of different affinity for substrate and effector nucleotides; these interactions will be discussed in a later paragraph. The most unusual features of the smaller subunit B 2 are its iron content and a tyrosine residue present as stable free radical. Both these components are essential for enzyme catalysis and structurally coupled. They confer characteristic light absorption to the protein, with a broad maximum around 370 nm (ε =

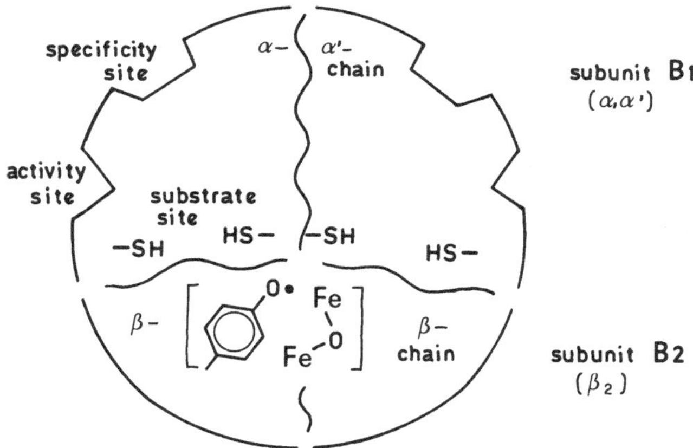

Fig. 2. Structure model of the ribonucleoside diphosphate reductase of *Escherichia coli*

8,000) and a sharp peak at 410 nm (ε = 6,000) besides the usual 280 nm-maximum; moreover a weak absorption band at 600 nm (ε = 500) has been identified in pure, concentrated samples[34, 35]. It is this specific spectrum that renders the reductase – unlike other proteins – very sensitive to photoinactivation by UV light and is in fact responsible for the lethal effect of near-ultraviolet (365 nm) irradiation on bacterial cultures[36].

The two atoms of iron per B 2 subunit are no heme-like porphyrin complexes, nor does the protein contain inorganic labile sulphide attached to the iron like in ferredoxins. Rather ribonucleotide reductase belongs to a small group of iron proteins where the metal is oxygen-coordinated, like in hemerythrins (see below). The iron can be removed from and reintroduced into enzyme subunit B 2 with concomitant activity loss and reactivation, respectively. Removal is difficult and not possible with simple chelating agents like EDTA but requires protein precipitation with acid ammonium sulfate or prolonged dialysis against a buffer containing 50 mM 8-hydroxyquinoline sulfonate, 30 mM hydroxylamine, and 1 M imidazole (pH = 7). The stable, metal-depleted apoprotein has no visible light absorption. It is best reconstituted by treatment with freshly prepared Fe(II)ascorbate solution under aerobic conditions[35, 37]. Fe(II) in the absence of oxygen, Fe(III)salts, or other metals are inefficient. Regeneration of the native structure obviously involves an oxidation step to Fe(III).

The nature of the iron center of *E. coli* ribonucleotide reductase is well documented by physical studies. It bears a remarkable, indeed surprising similarity with the structure of hemerythrins, a group of functionally unrelated, much smaller iron proteins. These oxygen-carrying polypeptides of marine worms contain Fe(III)-O-Fe(III) centers with histidine, aspartate and glutamate as amino acid side chain ligands and a nearby tyrosine residue[38]. Except for the 410 nm-band the electronic spectra of oxyhemerythrin and B 2 match closely, and both are diamagnetic at low temperature. Magnetic susceptibility measurements over the 30–200 K temperature range[35], Mößbauer spectra of reconstituted [^{57}Fe]protein B 2[37], and the analysis of Fe-O vibrations in the Raman sepctrum of native B 2 samples[39] are all consistent with a bent Fe-O-Fe structure of two non-identical, high spin Fe(III) ions antiferromagnetically coupled by a μ-oxo bridge. Comparison with model compounds like the pyridine-$[Cl_3Fe\text{-}O\text{-}FeCl_3]^{2-}$ complex[40], the observed exchange rate with solvent oxygen in $H_2^{18}O$ ($k = 8 \cdot 10^{-4}\ s^{-1}$), and the isotope shift of the Raman resonance ($15\ cm^{-1}$) indicate that the binuclear iron complex is only singly bridged and has a Fe-O-Fe angle of about 130°.

The abovementioned organic radical of subunit B 2 is easily recognized by an ESR signal consisting of a doublet around g = 2.0047 (Fig. 3)[41]. It disappears in iron-depleted apoprotein and is recovered upon iron reconstitution. On the other hand, the radical, the 410 nm absorption band, and enzyme activity are destroyed in parallel by treatment with hydroxylamine or hydroxyurea which do not affect the iron content or structure of protein B 2. (Note that hydroxyurea, NH_2–CO–NH–OH, does not act as an iron chelator here.) Thus the iron complex is necessary for generation of the radical but not vice versa; a stable radical in iron-free environment is probably not existent.

The nature of the free radical was established in a series of isotope substitution experiments in which reductase-overproducing bacteria grew in deuterated media with

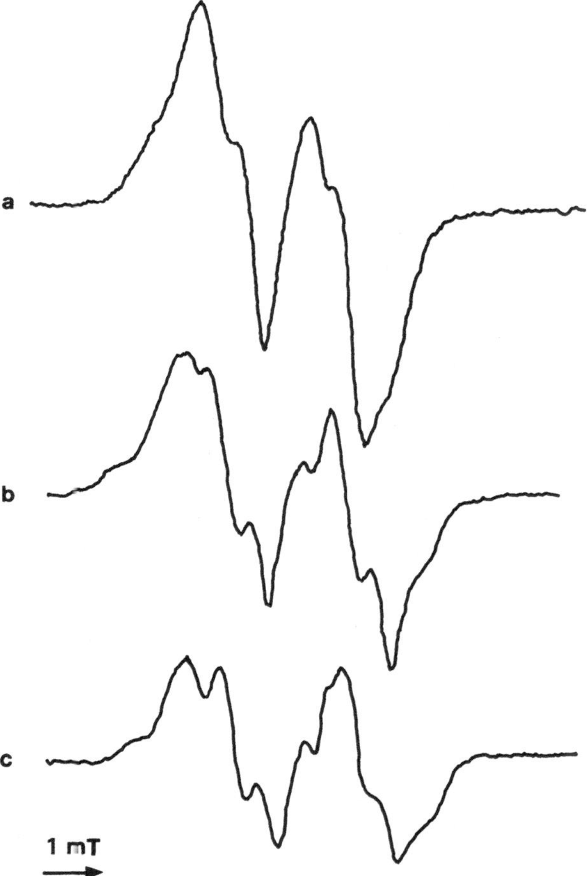

Fig. 3 a–c. ESR spectra of the tyrosine free radicals in ribonucleotide reductases. **a** protein subunit B 2 of *E. coli* enzyme, measured at 86 K[41]; **b** bacteriophage T 4 enzyme, at 77 K[90]; **c** hydroxyurea-resistant, reductase-overproducing mouse 3 T 6 cells, at 32 K[60]

individual non-deuterated amino acids. The drastic change of ESR spectrum after complete deuteration was reversed only by tyrosine; a tyrosine residue as the radical site was then verified by incorporation, in H_2O medium, of various deuterium- and ^{13}C-labelled tyrosines and analysis of the changes in hyperfine coupling[42, 43]. The radical is most likely an oxidized tyrosine, formed by (iron-catalyzed? oxygen-dependent?) electron abstraction from the aromatic ring, or an aryloxy species formed by H· abstraction or deprotonation of the simple aryl radical (scheme III); the ESR signal probably matches better the first structure while the 410 nm spectral band resembles that of phenoxy radicals. The unpaired electron must be delocalized over the phenyl ring, with an estimated spin density distribution as indicated. Comparative studies with bulky hydroxamates as radical scavengers[62] suggest that the tyrosine lies in a protein pocket; this may be one reason for the radical's stability which is unusual of an unsubstituted phenol derivative.

CH2 / OH —(−e)→ CH2 / 0.5, 0.26, 0.26 / OH ⇌(H+) CH2 / O• ↔ CH2 / O; −H• (III)

The free radical concentration does not significantly exceed 1 per B 2 subunit even in most active and most concentrated enzyme samples[35]. Thus, the molecular composition of this reductase subunit can be viewed as 2 tightly aggregated polypeptide chains, 2 coupled iron atoms, and 1 radical, or under a more functional aspect as 1 native protein containing 1 binuclear iron complex and 1 tyrosyl radical (Fig. 2).

In addition to these components and subunit B 1, *E. coli* ribonucleotide reductase requires Mg^{++} ions (in the 10 mM concentration range) for activity. Mg^{++} may be replaced by high concentrations (>0.2 M) of monovalent salt such as Na-acetate[32]. These ions serve no catalytic function but are essential for formation and integrity of the active B 1–B 2 complex.

The distribution of iron-dependent ribonucleotide reductase among other bacteria cannot be assessed at present. Unlike cobalt or manganese, iron deficiency in growth media would lead to various biochemical lesions difficult to distinguish from each other. No other bacterial enzymes have become known that contain non-heme-iron like the *E. coli* protein. However we shall return to Fe-ribonucleotide reductases in the viral and eukaryotic enzyme systems.

3. *Manganese Dependence*

The need for divalent manganese in cell proliferation of certain gram-positive bacteria has long been known but its involvement in ribonucleotide reduction was recognized only recently. Strains like *Brevibacterium ammoniagenes* or *Micrococcus luteus* cease to synthesize DNA (but not RNA or protein) in manganese-free fermentation media; the cultures accumulate ribonucleotides and are in fact used for industrial ribonucleotide production. Auling and Follmann could identify ribonucleotide reduction as the specific target of intracellular metal deficiency by the in vivo and in vitro correlations summarized in Table 1. It is seen that Mn^{++}-depleted cells which have very low DNA synthesis rate also lack ribonucleotide reductase activity, and that both can be restored by addition of Mn^{++} to cell cultures or to bacterial extracts, respectively[44]. In complete analogy to the overproduction of B 12-dependent reductase under cobalt deficiency in rhizobia[27], inactive apoprotein is formed in excess in manganese-depleted *Brevibacterium* cultures.

The ribonucleotide reductase of *B. ammoniagenes* was isolated from bacteria recovering from manganese starvation and was characterized in some detail[44, 45]. Like the Fe-reductases it is separated into inactive protein fractions on affinity chromatography materials such as Blue Sepharose and enzyme activity is restored after recombination of subunits. The enzyme is inactivated by dialysis vs. EDTA and can then be reactivated by metal addition in vitro (Fig. 4). In this process Fe^{++} can replace Mn^{++} with about half the efficiency of Mn^{++} whereas Co^{++}, Zn^{++}, or the cobalt coenzyme deoxyadenosylcobalamin are completely inactive. (However, Fe^{++} has no reactivating effect in Mn-depleted whole cells.) The situation resembles that with calf thymus ribonucleotide reductase in vitro where iron can be replaced by manganese ions[56]. Both observations may reflect the chemical similarity and isoelectronic nature of Mn^{++} and Fe^{+++} that result in similar interactions with isolated apoenzymes, whereas the metals could well be discriminated against each other under physiologic conditions of metalloenzyme formation. It may further be noted that both metals share the capacity to form binuclear μ-oxo complexes, and that both enzymes are efficiently inhibited by hydroxyurea.

Additional support for the Mn^{++} specificity of native *B. ammoniagenes* ribonucleotide reductase is provided by its inhibition by cyanide (Fig. 5). The iron- and B 12-enzymes are not affected by cyanide which forms cyano and cyanoaquo complexes with Mn^{++} but not with Fe^{++} or Fe^{+++} under typical enzyme assay conditions at ambient temperature. We thus believe that manganese in the *Brevibacterium* enzyme system is the functional, catalytic counterpart of iron found in *E. coli* and eukaryotic ribonucleotide reductases. The nature of Mn^{++}-protein interaction remains to be studied after further purification.

Table 1. Manganese dependence of DNA synthesis (adenine incorporation in vivo) and ribonucleotide reductase activity in *Brevibacterium ammoniagenes*[44]

Culture conditions	DNA synthesis rate (%)	addition to enzyme assay	reductase activity (nmol GTP $\cdot$ h^{-1} $\cdot$ mg^{-1})
Mn present	100	none	0.9
		10 mM Mn^{++}, Mg^{++}	2.1
Mn-depleted medium ("0-cells")	15	none	0.1
		10 mM Mn^{++}	1.8
		10 mM Fe^{++}	0.9
		10 mM Mg^{++}	0.2
0 cells in fresh medium (Mn present)	230	none	2.9
		10 mM Mn^{++}, Mg^{++}	6.6

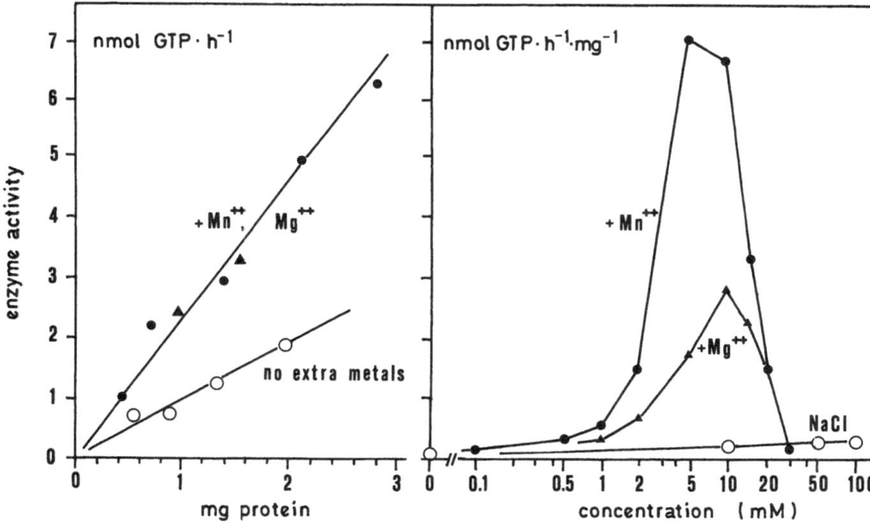

Fig. 4. Stimulation and reactivation of *Brevibacterium ammoniagenes* ribonucleotide reductase by metals. *Left:* Untreated enzyme extract; Mg^{++} and Mn^{++} are equally active. *Right:* Mn^{++}-specific reactivation of enzyme solution inactivated by 4 h dialysis vs. EDTA; removal of Mn^{++} is incomplete under these conditions, explaining the residual activation by Mg^{++}. More prolonged dialysis denatures the enzyme

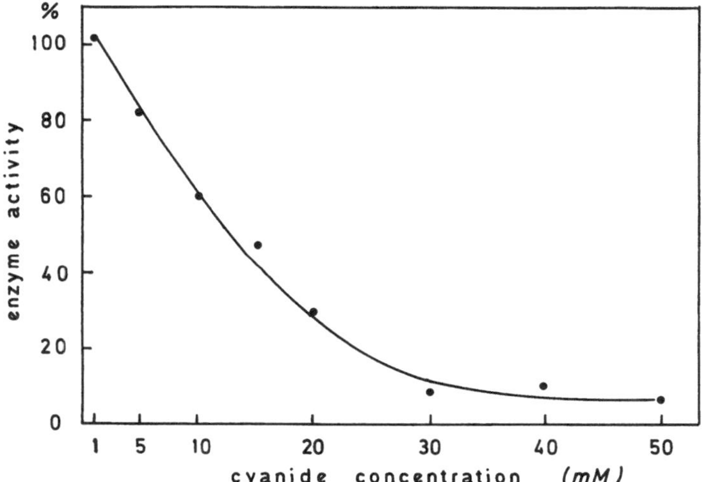

Fig. 5. Inhibition of *Brevibacterium ammoniagenes* ribonucleotide reductase by cyanide. Conditions: GTP as substrate, 0.5 mM dTTP as effector; 2 mM dithiothreitol; pH 7, 30 °C, 0.1 M NaCl

Structure analysis of the new enzyme is complicated by its general preference for high ionic strength (0.2 M NaCl) and for divalent cations; both may reflect a physiologic status of the bacteria which are usually grown in fermentation media of high salt content. Mg^{++} or Ca^{++} stimulate the activity of non-EDTA-treated, native enzyme samples. In Mn^{++}-

depleted preparations this is hard to differentiate from the Mn^{++} reactivation effect because Mg^{++} and Mn^{++} are well known for their exchangeability in binding to biomolecules. We assume that *B. ammoniagenes* ribonucleotide reductase needs divalent cations for structural integrity like the *E. coli* or *Anabaena* enzymes besides the specific Mn^{++} requirement for catalysis.

The distribution of manganese-dependent DNA synthesis and ribonucleotide reduction is currently being studied by Auling[46]. Because there is no simple assay for this type of reaction in bacterial extracts one has to follow the Mn^{++} dependence of DNA labeling in vivo and only in positive cases enzyme activity is then characterized in vitro. The five species of gram-positive bacteria identified in this way are included in Table 4.

4. *Eukaryotic Ribonucleotide Reductases*

Characterization of the ribonucleotide reductases in eukaryotes has lagged behind knowledge of the bacterial enzymes for many years, mostly because of the much slower growth rates of higher plants and animals. Unicellular organisms like algae or yeast, seedlings, embryonic or regenerating organs, or tumor cells are suitable sources for enzyme purification whereas it is almost impossible to measure ribonucleotide reduction in adult plant or animal tissues. Remarkable similarities exist between the ribonucleoside diphosphate reductases of green algae and of mammals, e.g. in calf thymus, suggesting that the majority of eukaryotes shares one type of enzyme which is distantly related with the iron reductase of *Escherichia coli.* To date only two examples of B 12-dependent ribonucleotide reduction have been found in the eukaryotic "urkingdom", *viz.* in *Euglena gracilis*, discussed above[24–26], and in a fungus *Pithomyces chartarum*[47] where the enzyme has not been characterized in detail. Mammals, including man, of course depend on dietary vitamin B 12 uptake but despite several efforts deoxyadenosylcobalamin involvement in ribonucleotide reduction as the cause of DNA synthesis distortions in anemic cells could never be demonstrated.

Deoxyribonucleotide biosynthesis in lower eukaryotes is being studied in our laboratory. Synchronous cultures of the unicellular freshwater algae, *Scenedesmus obliquus* overproduce ribonucleotide reductase more than hundredfold when DNA synthesis is blocked by 5-fluorodeoxyuridine, and such cells serve as a good source of enzyme[48]. Gel permeation chromatography yields an enzyme fraction of M_r = 250,000 capable to reduce pyrimidine and purine ribonucleoside diphosphates in presence of dithiothreitol or thioredoxin and an allosteric effector (ATP, or dTTP). The active holoenzyme easily dissociates into two non-identical, inactive subunits, designated U 1 and U 2, during affinity chromatography on immobilized nucleotide ligands or even in sucrose density gradients; recombination of U 1 and U 2 then restores enzyme activity. U 1 sediments as a protein of M_r = 170,000 (8.5 S). It is obtained in apparent homogeneity when algal proteins are chromatographed on a reductase-specific affinity adsorbent, aminohexyl-dATP Sepharose[49]; while U 2 is not retarded, U 1 is tightly bound and specifically eluted by 1 mM dATP, resulting in 4,500-fold purification. This material exhibits a molecular mass of 90,000 on SDS polyacrylamide gels indicating that the native nucleotide-binding subunit is a dimer. U 2 is much more difficult to purify; our most active fractions behave as one protein of $M_r \approx$ 75,000 in the ultracentrifuge and on SDS gels without further dissociation.

The algal enzyme does not require added Mg^{++} or other ions for activity in vitro; concentrations of more than 0.1–1 mM Fe^{++}, Mn^{++}, or Mg^{++} are in fact inhibitory. However it is identified as an iron enzyme by the inactivating effect of chelating agents like hydroxamic acid analogs, hydroxyurea, or 1,10-phenanthroline and by subsequent reactivation in presence of ferrous salts (Fig. 6 and 7).

Synchronous or exponentially growing cultures of baker's yeast (*Saccharomyces cerevisiae*) are another convenient source of ribonucleotide reductase of good activity and yield[50–52]. The enzyme resembles the algal reductase in its ready dissociation into a nucleotide-binding and a catalytic subunit, independence of exogenous metal ions, insensitivity towards cyanide and millimolar concentrations of EDTA, and specific inactivation by hydroxyurea.

It has so far been impossible to isolate ribonucleotide reductases from homogenates of higher plants such as the germinating wheat embryo, or bean root tips and cell cultures which all contain only minute amounts of enzyme[53,54]. As these activities are clearly growth-correlated like in other organisms and standard assay conditions are required, there is no evidence for a plant-specific, diverse type of reductase.

Mammalian ribonucleotide reductase was first isolated from the fast-proliferating Novikoff rat tumor[55] but thymus glands from young calves appear to be the best tissue for enzyme preparation in milligram quantities[56–58]. Two non-identical proteins M1 and M2, inactive separately, were obtained and were shown to consist of polypeptide chains

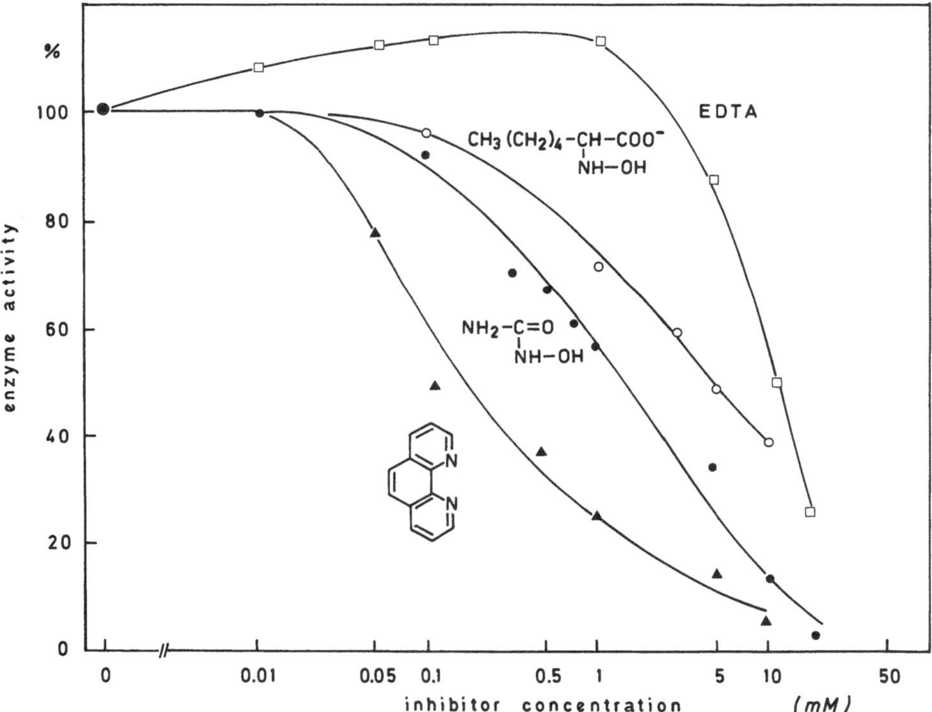

Fig. 6. Inactivation of ribonucleotide reductase from the green alga, *Scenedesmus obliquus* by various chelating agents

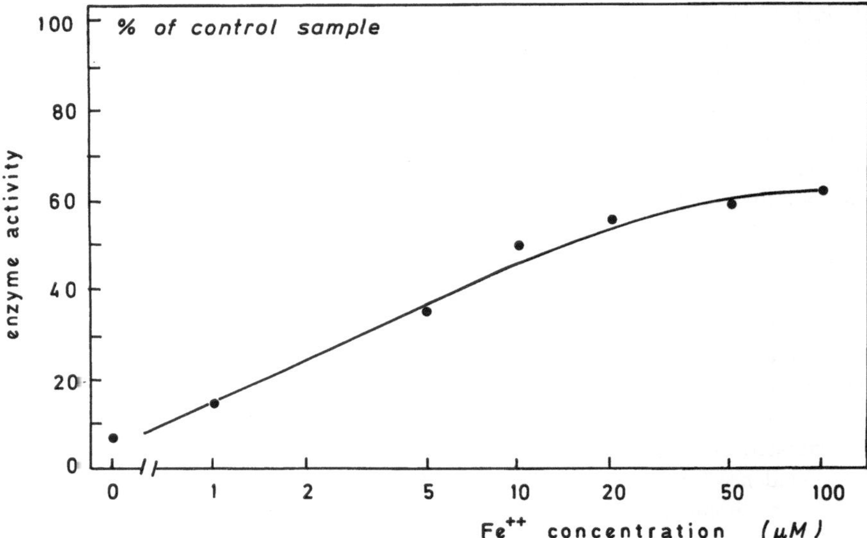

Fig. 7. Reactivation of *Scenedesmus obliquus* ribonucleotide reductase by Fe^{++} ions following 1 h treatment with 10 mM hydroxyurea and 3 hrs of dialysis

of 84,000 and 58,000 dalton, respectively. Homogenous M 1 binds nucleotides and in the presence of thymidine triphosphate it is a dimer (M_r = 170,000); tetramers are formed upon binding of dATP. M2 is also a dimer under non-denaturing conditions (M_r = 110,000) but is more difficult to purify to homogeneity. Enzyme activity of the combined subunits is twofold stimulated by magnesium but not by iron ions; it is inhibited by EDTA Reactivation is then possible by addition of Fe(II)salts or $MnCl_2$, but not zinc or cobalt salts. 1 mM hydroxyurea completely inactivates calf thymus ribonucleotide reductase but its action is reversed by simple removal of the compound on a Sephadex column without necessity to replenish metals. Radioactive ^{59}Fe was used to label the protein[56]. The metal must be bound to M2 which, in contrast to M1, has an absorption spectrum with bands at 325 and 404 nm[58].

A characteristic property of calf thymus enzyme preparations is the low, substoichiometric amount of M2 (approximately 1/10 of M1). However it is possible to study M2 in more detail by combining M1 with an M2 preparation isolated from hydroxyurea-resistant mouse fibroblast 3T6 cells which overproduce protein M2[59, 60]. In this system one observes a low temperature ESR signal at g = 2 resembling the signals given by the tyrosine radicals of the B2 subunit of *E. coli* and phage T4 ribonucleotide reductase (Fig. 3). It was again identified as a tyrosine radical, with the unpaired electron delocalized over the aromatic ring, by growing the mouse cells in presence of specifically deuterated amino acids, [β,β-$^2H_2$]- and [3,5-$^2H_2$]tyrosine, respectively. Substantial differences in the radical centers of mammalian and bacterial enzyme are observed in the ability to regenerate the free radical after destruction with hydroxyurea. Radical content of M2 and enzyme activity of the holoenzyme reappear fast when hydroxyurea is removed and when samples containing dithiothreitol or thioredoxin plus iron ions are kept under aerobic conditions, resembling the easy reactivation of *Scenedesmus* ribonucleotide reductase following hydroxyurea treatment.

The role of oxygen in eukaryotic DNA biosynthesis may indeed be a critical one. It has recently been shown that O_2 is not only required for initial formation of tyrosyl radical but must be continously present to maintain the radical content and enzyme activity of mammalian ribonucleotide reductase[346]. In vivo studies with Ehrlich ascites cells also point to a tight link between oxygen and deoxyribonucleotide supply: Anaerobic arrest of cells in G1 phase and block of DNA synthesis can be relieved by addition of deoxycytidine, but not cytidine, to the culture medium[347].

The topology of the radical site pocket of calf thymus and mouse fibroblast ribonucleotide reductase was recently probed with a series of hydroxamate inhibitors of increasing bulkiness[62] and will be discussed in the following section. Other mammalian sources from which ribonucleotide reductases have been isolated and more or less purified include: rat Novikoff hepatoma[55, 61] and regenerating rat liver[63, 64]; rabbit bone marrow[65, 66]; Ehrlich ascites tumor cells of mice[67–69]; and cultured human lymphoblast cells[70]. Some of their properties are described in Table 2. Many more animal and human cells were assayed for enzyme activity, frequently in mutant cell lines, to test for cell cycle dependence, mechanisms of metabolic regulation, drug resistance, and correlation with tumor growth rates. Representative studies of this kind, which rapidly expand in number, are summarized in Table 3.

From Table 2 it may be seen that the subunit composition and iron requirement of all eukaryotic ribonucleotide reductases are basically similar: one larger, nucleotide-binding protein always has to combine with another (usually smaller) one, probably catalytic in nature. (An agreement on a reasonable denomination scheme for these subunits would appear timely.) It is difficult at present to assess the range of molecular weights reported for the constituent polypeptide chains. Not only have they been determined using different techniques (by gel filtration, SDS gel electrophoresis, sedimentation equilibrium, or velocity methods) but their mode of aggregation also depends on presence or absence of effector nucleotides (ATP, dATP, or dTTP), which have not been systematically analyzed in each individual case. The more intriguing possibility – for which there is increasing evidence – is that the eukaryotic enzymes have some inherent flexibility to form polypeptide chain complexes of varying composition (e.g., 2:1, 1:1, or 1:2) depending on the bound substrate and effector nucleotides, and serving a functional role through the kinetic changes that result[168]. The non-stoichiometric ratio of subunits[56] and their non-coordinate changes during growth of rodent cells[69, 71] are compatible with such a view. (In contrast, *E. coli* ribonucleotide reductase subunits change in coordinate fashion[72].) Purification of the rat liver enzyme[64] has even led to the suggestion that four different L1 subunits exist, each specific for reduction of one of the four ribonucleotide substrates when attached to the L2 subunit; however this is in conflict with the capacity of pure calf thymus reductase to reduce all ribonucleotides alike. Another complication, to be discussed below, is that ribonucleotide reductases intracellularly are part of a large multienzyme cluster[83] where the mode of interaction and the function of their subunits may be different again. In conclusion, a molecular structure of *the* eukaryotic ribonucleotide reductase cannot yet be drawn. Nevertheless it is not unreasonable to visualize that the more sophisticated regulatory demands on genome replication in eukaryotes reflect themselves in an enzyme system that is even more modulatable than the bacterial proteins, which are also under effector control but at least possess fixed stoichiometry of subunits.

Table 2. Structure of purified eukaryotic ribonucleotide reductases. Data of *E. coli*- and T4-enzymes are included for comparison

Source	Protein subunits	Nucleotide-binding subunit	Total mol. mass (kdalton)	Mass of component polypeptides (kdalton)	Evidence for iron	Absolute requirement for Mg^{++}	Approx. purification factor	Ref.
green algae (*Scenedesmus obliquus*)	U 1 + U 2	U 1	250	(2 · 90) + 75	yes	no	4500	48
yeast (*Saccharomyces cerevisiae*)	U 1 + U 2	U 1			?	no		52
calf thymus	M 1 + M 2	M 1	280	(2 · 84) + (2 · 55)	yes	no	3400	57, 58
Novikoff hepatoma (rat)	P 1 + P 2	P 1	250	(2 · 90) + ?	yes	(yes)	700	61
regenerating rat liver	L 1 + L 2	L 1	280	(4 · 45) + (45 + 75)[a]	yes	no		63, 64
rabbit bone marrow	S 1 + S 2	(not separated)			yes	?	6000	66
Ehrlich ascites tumor (mouse)	"dye" + "Tris"	"dye"	250–300	m(95 to 127) + n(75 to 81)[a]	yes	?		68
human lymphoblasts (Molt 4 F cells)	B + A	B	210	100 + 100	?	(yes)	400	70
Escherichia coli	B 1 + B 2	B 1	245	(2 · 80) + (2 · 39)	yes	yes	1500	33
bacteriophage T 4	$\alpha_2 + \beta_2$ (B 1 + B 2)	B 1; difficult to separate	225	(2 · 85) + (2 · 35)	yes	no	150	90

[a] Complex data, *cf.* discussion in the text and references

Table 3. Studies of cell cycle dependence and properties of ribonucleotide reductase in mammalian cells

Source	Ref.
Mouse L cells	73–75
Hydroxyurea-resistant mouse cells	59, 76
Leukemic mouse spleen	77
Thymidine-resistant mouse fibroblasts	78
Mutant mouse lymphoma cells	79
Numerous rat hepatomas and other tumors	80, 81
Cultured hamster fibroblasts	82, 83, 244, 320
Developing brain (mouse, rat, human)	84
HeLa cells	85
Phytohemagglutinin-stimulated human lymphocytes	86
Various human cell lines	87
Intact human diploid fibroblasts	88

5. *Viral Ribonucleotide Reductases*

When a virus infects a cell, it could induce its host to make deoxyribonucleotides for viral DNA replication by means of the cellular enzyme (probably derepressed after infection), or the virus could carry its own specific ribonucleotide reductase genes which are expressed in the host cell and produce a new enzyme. Ribonucleotide reductase of herpes simplex virus is even considered the transforming function of HSV-2[348]. However virus-specific reductases are not easy to identify unless one is able to differentiate between the host cell and viral enzymes. The only two systems where this has been unambiguously achieved concern, therefore, bacteriophage T replication in *Escherichia coli*, and infection of mammalian cells by members of the herpes-virus group.

Phage T4 ribonucleoside diphosphate reductase[89–92] is similar to the *E. coli* enzyme in general structure. It consists of two non-identical subunits B1 and B2 each made up from two polypeptide chains ($\alpha_2\beta_2$). Molecular weights were determined to 225,000 for the holoenzyme and about 85,000 and 35,000 for the α and β chains, respectively. These proteins are products of phage genes *nrdA* and *nrdB* and they do not form catalytically functional hybrids with *E. coli* reductase. They also differ from bacterial enzyme in that B1 and B2 do not dissociate during purification; accordingly magnesium ions are not essential for activity. Subunit B2 of the phage enzyme contains 2 iron atoms which are removed in 1 M guanidinium chloride solution (under inactivation) and are easily reintroduced with Fe^{++} salts[90]. The same subunit harbours a tyrosine free radical which gives rise to an ESR doublet signal centered at g = 2.0049 (Fig. 3)[92]; it is similar but not quite identical to that of *E. coli* reductase. The T4 enzyme also differs in its sensitivity to hydroxyurea, being about an order of magnitude more susceptible than its bacterial counterpart[62, 91].

Bacteriophages T5 and T6, but not T7 or a λ phage, also induce new ribonucleotide reductase activities in infected *E. coli* cells[93]. The enzymes of T4 and T6 were immunologically related but did not cross-react with *E. coli* enzyme. The unstable reductase of phage T5 differed again; because of its hydroxyurea sensitivity it is probably an iron protein, too.

Mouse, hamster, and human cell lines have been infected with Herpes simplex, Epstein-Barr virus (EBV), or pseudorabies virus and a strong increase in ribonucleotide reductase activities was always observed[94–98]. These enzymes usually differ from the cellular reductases in lower sensitivity to Mg^{++} and other ion concentrations and in lack of inhibition by the common negative effector nucleotides, dATP or dTTP. Hydroxyurea resistance was reported in case of EBV infection[95]. The most convincing evidence for a virus-specific reductase was recently obtained by Thelander[98] who showed that the pseudorabies virus-induced enzyme of mouse L cells is not influenced by antibodies against calf thymus ribonucleotide reductase whereas the activity in uninfected cells was completely neutralized. The new enzyme showed an ESR signal resembling, but distinct from that of the tyrosine radical of the M2 subunit of mammalian reductase, and it had the same sensitivity towards inhibition by hydroxyurea. This viral ribonucleotide reductase is possibly not regulated at all by effector nucleotides and may prove a particularly simple one, making its further structure analysis a valuable effort.

6. Ribonucleotide Reduction: A Universal Reaction

As stated above in Scheme II, there is no route other than ribonucleotide reduction that could provide normal proliferating cells with the necessary DNA precursors. Only in few cases can, for a certain period of time, DNA be made without ribonucleotide reduction, viz. in microorganisms or cultured higher cells that salvage 2-deoxyribose or deoxyribonucleosides from DNA or deoxyribonucleotide degradation, and in specialized tissues like sea urchin eggs or plant seeds where deoxyribotides are stored for the developing embryo[53, 99]. The normal fate of DNA after cell death or ingestion by another organism is complete destruction, including cleavage of the deoxysugar to acetaldehyde and glyceraldehydephosphate by deoxyribosephosphate aldolase. Ribonucleotide reduction must therefore be a universal process of life just like ribosomal protein synthesis or the RNA and DNA polymerase reactions. In fact it must have preceded the emergence of DNA on the way from protobionts to "true", DNA-containing early cells[10].

An attempt is made in Table 4 to list all prokaryotic and eukaryotic organisms in which ribonucleotide reductase activity has been demonstrated. It is not only meant to prove the point of universality but even more to draw attention to obvious gaps where better knowledge would be desirable. The existence of three different metalloenzyme systems for catalysis of the same reaction raises phylogenetic questions which cannot be answered from the table as the reaction remains to be characterized in large groups of bacteria (e.g., the purple or green photosynthetic bacteria), in most fungi, or protozoa.

At present one may state that iron dependence of ribonucleotide reduction is typical for eukaryotes and B 12-dependence is widely distributed among bacteria, but with numerous exceptions. Screening for manganese dependence has only recently begun. The iron reductase of *E. coli* is presently the only bacterial enzyme related to the eukaryotic proteins but other enterobacteriaceae have not been investigated. Not too uncommon is the observation that among genera of bacteria with B 12-dependent ribonucleotide reductases there are individual strains that do not utilize the cobalt coenzyme, for example among *Bacillus, Micrococcus, Pseudomonas*, or cyanobacterial (*Synechococcus*) species, but the actual metal requirement is frequently unknown altogether. In classifying the type of ribonucleotide reduction in a certain organism, its sensitivity, or

Table 4. Experimental evidence for ribonucleotide reduction and its metal (coenzyme) requirement in various groups of organisms

Organism	Type of enzyme			Hydroxyurea sensitive	Ref.
	Fe	Co (B 12)	Mn		
Prokaryotes					
Bacteria (Eubacteria)					
Clostridium sticklandii, C. tetanomorphum, C. thermoaceticum		■			18
Lactobacillus leichmannii, L. casei, L. acidophilus		■			11, 16, 18
Bacillus subtilis				yes	100, 101
Bacillus megaterium		■			102
Streptococcus faecalis				yes	101
Propionibacterium shermanii		■			103
Corynebacterium nephridii		■			18, 23
Streptomyces aureofaciens		■			104
Brevibacterium ammoniagenes		–	■	yes	44
Arthrobacter globiformis, A. citreus			■	yes	46
Nocardia opaca			■	yes	46
Micrococcus luteus		–	■	yes	44
Micrococcus (Paracoccus) denitrificans		■			18
Pseudomonas stutzeri		■			18
Pseudomonas aeruginosa		–		yes	18, 119
Escherichia coli	■	–		yes	2, 3, 18, 32
Klebsiella sp.				yes	101
Salmonella typhimurium					120
Thermus aquaticus (X–1)		■			18, 21
Rhizobium meliloti, R. trifolii, R. phaseoli, R. japonicum		■			27, 28
Cyanobacteria					
Anabaena 7119		■			22
Anacystis nidulans		■			19, 20
Nostoc commune		■			19, 20
Synechococcus cedrorum and 10 more species		■			19, 20
Agmenellum quadruplicatum		–		yes	20, 105
Coccochloris elabens		–			20
Archaebacteria					
Methanobacterium thermoautotrophicum		–			106

Table 4 (continued)

Organism	Type of enzyme			Hydroxyurea sensitive	Ref.
	Fe	Co (B 12)	Mn		
Eukaryotes					
Fungi					
Pithomyces chartarum		■			47
yeast (Saccharomyces cerevisiae)		–		yes	50, 52
Achlya sp.				yes	107
Plants					
Euglena gracilis		■		yes[a]	24, 121, 122
green algae (Scenedesmus obliquus, Chlorella pyrenoidosa)	■	–		yes	48, 108
wheat embryo (Triticum aestivum)		–			53
broad bean (Vicia faba) root tips, soy bean (Glycine max) cells		–		yes	54, 123
Animals					
sea urchin eggs (Arbacia punctulata, Paracentrotus lividus)				yes	109, 110, 124
silk moth pupae (Hyalophora cecropia)				yes	111
amphibian eggs (Xenopus laevis, Pleurodeles waltlii)				yes	112, 124
chick embryo				yes	113, 114
rat, mouse, hamster, rabbit tissues, cells and tumors	■			yes	55, 63–67, 73–84, 244
calf thymus	■			yes	56
rhesus monkey tumor					115
human cells	■	–		yes	70, 85–88
Viruses					
bacteriophages T4, T5, T6 in E. coli	■			yes	89, 90, 93
Herpes viruses in mammalian cells	■			yes[b]	94–98

[a] inhibits chlorophyll synthesis (121); [b] *cf.* text (p. 46)

resistence against hydroxyurea can be helpful: Sensitivity to the drug usually rules out B 12-dependent deoxyribonucleotide formation as the B 12 enzymes are resistant to hydroxyurea in vitro. In contrast hydroxyurea sensitivity does not suffice to differentiate between iron and manganese dependence of DNA synthesis (cf. Sect. D.2).

Table 4 contains one organism from the remote group, or "third urkingdom"[118] of archaebacteria. Evidence for the reductive pathway of deoxyribonucleotide formation was obtained by in vivo DNA labeling of the methanogenic bacterium *M. thermoautotrophicum*, strain Marburg, using the ribonucleoside cytidine which has to enter DNA via reduction[106]. However the reaction was not measurable in cell-free extracts of the strict anaerobes, probably because the proper redox state, metal or cofactor requirement had not been met in vitro. Deoxyadenosylcobalamin had no effect. In view of the unique, still not fully explored cofactor array of the methanogens, including a nickel tetrapyrrol[116], it is tempting to speculate whether they have yet another ribonucleotide reductase system.

The hypothesis has been advanced that B 12-dependent ribonucleotide reduction is a characteristic of anaerobic, possibly older and more primitive microorganisms like clostridia and lactobacilli[117]. It would be consistent with the fact that the iron reductases of *E. coli* or mammals obviously require oxygen for radical generation. While there may be a correlation of this kind in many instances the classification cannot be a general one in view of exceptions observed with the methanogens or facultative anaerobes like cyanobacteria, *E. coli*, or yeast which make DNA independent of cobalamins. The intriguing possibility that cells might employ two different enzyme system under anaerobic or aerobic conditions has not been explored. A brief report of such a situation in vitamin B 12-producing *Propionibacterium shermanii*[103] was not investigated further.

We conclude that enzyme catalysis of deoxyribonucleotide formation is a universal but at the same time one of the most diversified biochemical processes; on the other hand mechanistic aspects which tend to unify the picture will be discussed in the next section. An attractive goal is to compare this particular group of enzymes with the natural ancestry of all living organisms established by Woese[118] but their phylogeny is not within easy reach for the difficulties described in Sect. C.4. It has to remain open at present whether the emerging correlation of ribonucleotide reductases with the main branches of the new natural system reflects the intracellular availability of catalytic metal ions, families of related enzyme proteins, or both.

C. Mechanisms and Structures

Understanding the mechanism of substitution of a ribonucleotide's 2′ OH group by hydrogen (Scheme I) is still a challenge for organic, physical, bioinorganic, and biological chemists; is it a (crypto)ionic or a radical process? Understanding how an enzyme can be specific for ribonucleotides yet unspecific enough to reduce the four of them at comparable rates, or how the characteristic modulation of substrate specificity by effector nucleotides works are puzzling questions for the enzymologist; or are there separate catalysts for reduction of adenine, cytosine, guanine, and uracil nucleotides? Unfortunately most mechanistic studies discussed in this section have been done only with the pure enzymes of *Lactobacillus leichmannii* and *Escherichia coli*. Mechanisms and specificities do not appear to differ grossly in the other purified enzyme systems in vitro but there could still be significant individual variations in eukaryotes which cannot be evaluated at this time.

1. Specificity for Ribonucleotides and Hydrogen Donors

Both natural substrates of a ribonucleotide reductase, ribonucleoside 5′-phosphates and dithiols (Scheme I) are complex molecules composed of several substructures. The enzymes share an interesting specificity pattern in tolerating manifold (although not unlimited) modifications at all but one substructure which must not be altered (Fig. 8).

At the ribose 5′-phosphate end, a diphosphate or triphosphate is always required for substrate property; reduction of 5′-monophosphates does not occur. Most enzymes are diphosphate reductases and fewer prefer the triphosphate level, e.g. the B 12 enzymes of

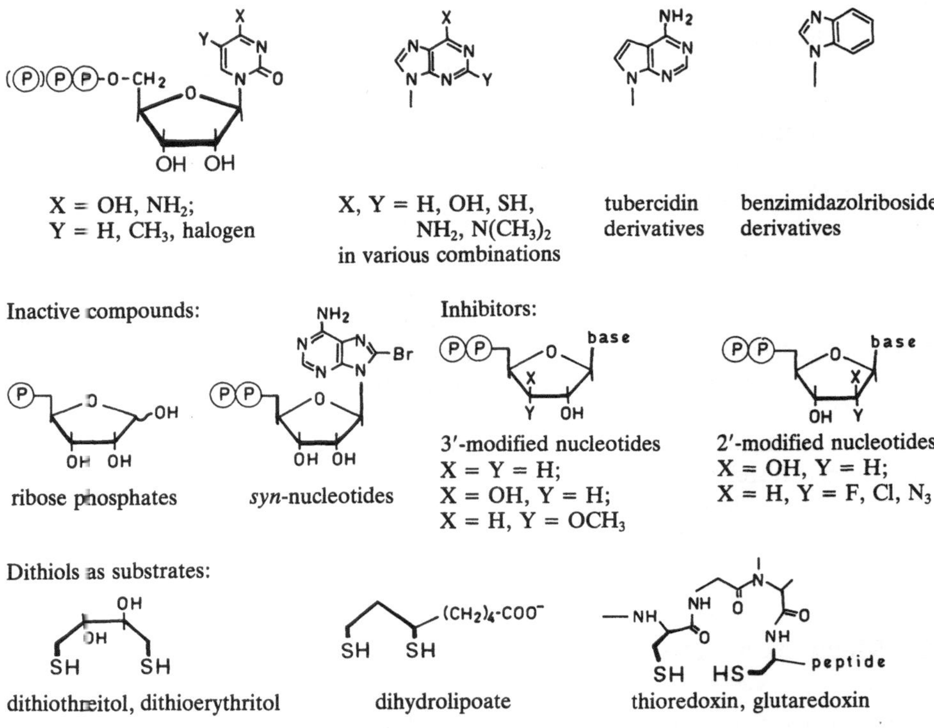

Fig. 8. Structure specificity of ribonucleotide reductases. P = phosphate residues

L. leichmannii and *Euglena gracilis*. Some appear to accept both types of ribonucleotide in vitro, for example the reductase of *Rhizobium meliloti*[28]. This specificity is not easy to establish because accompanying diphosphokinase activities are frequently present. 5′-Thiodiphosphates are also reduced by *E. coli* ribonucleotide reductase[125]. Substrate-protein interaction of the phosphates with an equivalent number of positively charged amino acid residues, or mediated by divalent metal ions, is certainly prerequisite for the catalytic process but probably not further involved in it.

It is the ribose moiety that unambiguously specifies a nucleotide as ribonucleotide reductase substrate. A 6′-diphosphate of 5′-deoxyallofuranosylcytosine ("homo-CDP") having an extra CH_2 group between the phosphates and the furanose ring is inactive towards rat tumor reductase[140]. A pentose 2′-OH group alone does not suffice either but must be part of D-ribose unmodified at any of the 2′- or 3′-positions. Thus, 2′- and 3′-O-methylribonucleotides, xylo- or 3′-deoxyribonucleotides are not reduced, nor are they inhibitory[126]. Synthetic nucleotides carrying modifications at C2′ (Fig. 8) form a series of inhibitors. Diastereomeric arabinonucleotides in which the inversed 2′-OH group is not reduced show weak, competitive or mixed-type inhibition[127]. Reactive nucleotides, functionalized at C2′ like in 2′-deoxy-2′-chloro-UDP, 2′-deoxy-2′-azido-CDP[128], or 2′-fluoro-nucleotides[129] make efficient "suicide" or k_{cat} inhibitors. The chloro derivative irreversibly modifies SH groups of the B1 protein of *E. coli* while it is decomposed to uracil, chloride and pyrophosphate ion, and an unidentified sugar fragment[129]. On the other hand the azido group of the CDP analog reacts with, and inactivates, subunit B 2 without affecting B 1; the nucleotide decomposes to cytosine. In this reaction which may resemble substrate reduction itself the radical of B2 is destroyed[130]. 2′-Deoxy-2′-azidonucleotides also inactivate *Lactobacillus* and mammalian ribonucleotide reductase[128].

Whether or not pyrimidine and purine ribonucleotides are reduced by a single enzyme is a matter of some controversy. The monomeric reductase of *L. leichmannii* and the enzyme of *E. coli* where substrates bind exclusively to B 1[131] as well as calf thymus and human reductase[132, 133] consist of only one holoenzyme species capable to reduce all the ribonucleotides. In contrast rat liver ribonucleotide reductase during purification dissociated into one protein fraction specific for CDP reduction and one reducing the three other diphosphates from which a specific UDP reductase activity could then be obtained[64]. Other reports of differential ADP and CDP reductase activities in rodent enzyme preparations[82, 134] are difficult to assess. In reductase assays one can easily encounter slow reduction rates for one particular ribonucleotide when not all conditions like pH, ionic strength, effector nucleotide and dithiol concentrations, and stoichiometry of subunits, are at an optimum. Therefore differential activities not necessarily indicate different enzyme species.

A useful though laborious alternative to demonstrate specificity for all natural substrates is to systematically study base-modified ribonucleotides. If there are separate entities, each should be highly specific for just one base structure; if intermediate nucleotides were also reduced this would argue for one substrate site of broad specificity. Several studies[126, 135–137] done with a great number of synthetic or natural nucleotides unambiguously confirm the latter situation for the bacterial enzymes; in contrast to their absolute ribotide specificity they will reduce almost all base-modified compounds (Table 5). With few exceptions (e.g. reduction of 5-fluoro-UDP by ascites cell reductase[138]) such data are not yet available for eukaryotic ribonucleotide reductases.

Table 5 contains ribonucleotides with the common amino and carbonyl structures, with extra substituents, and with totally unsubstituted bases like purine or benzimidazole. Apparent K_m values and velocities do not vary more than about tenfold in the presence of specific effector deoxyribonucleotides. Guanine and cytosine nucleotides have usually fastest and compounds with fewer base substituents show decreased reaction rates. Loss of substrate activity is only observed in *syn*-oriented nucleotides where the nucleobase is rotated about the glycosidic bond like in 8-bromo-ADP or -ATP. Molecular conformation-enzyme activity relationships have been discussed in detail[136].

Looking at a substrate site (in combination with an adjacent effector nucleotide site discussed below) we summarize the available information on molecular interactions as depicted in Fig. 9:

- Neutralisation of the 5′-phosphate anions, most likely by lysine and arginine residues;
- Concerted binding at the unsubstituted 3′ and 2′ hydroxyl groups of ribose, probably through hydrogen bridges. This is derived from the inactivity of sugar-modified nucleotides[126] indicating that removal of 2′ OH cannot occur without participation of 3′–C–OH, nor interaction at 3′ is possible without concomitant reduction at 2′.
- A flexible, two-sided induced fit of protein residues around the nucleobases in either nucleotide site[135]. Hydrogen bridge-forming base substituents like $-NH_2$ and $=O$ must play a prominent role in nucleotide-amino acid side chain interactions, judging from

Table 5. Kinetic data[a] and effector specificity for reduction of natural and synthetic, or modified ribonucleotides catalyzed by the triphosphate reductase of *Lactobacillus leichmannii* and diphosphate reductase of *E. coli*, respectively. (Data from Ref. 135, 136, 142)

Substrate: Tri- or diphosphate of	*Lactobacillus* enzyme				*E. coli* enzyme			
	Specific effector	Relative rate	$K_{M(app.)}$ (mM)	V_{max} (nkat)	Specific effector	Relative rate	$K_{M(app.)}$ (mM)	V_{max} (nkat)
guanosine		100	0.2		dTTP	100	0.05	0.08
inosine	dTTP	38			dTTP	75	0.11	0.06
adenosine	dGTP	100	0.22	0.40	dGTP	40	0.09	0.03
N^6-dimethyladenosine	dGTP	5	0.35	0.04	dGTP	slow		slow
2-aminopurineriboside					dTTP	70	0.04	0.05
2,6-diaminopurineriboside					dTTP	56	0.04	0.04
8-bromoadenosine		0		0	–	0		0
purineriboside	dGTP	19	0.67	0.17	dTTP	44	0.43	0.04
tubercidin	dGTP	75			dGTP	14		
benzimidazoleriboside	dGTP	2		slow	dTTP	48	0.07	0.04
cytidine	dATP	100	0.13		dTTP, ATP	82	0.10	0.07
uridine	dCTP	37			ATP, dTTP, dCTP	40		0.02

[a] measured spectrophotometrically at pH 7.7, 25 °C, in presence of thioredoxin/thioredoxin reductase/0.4 mM NADPH, and 10 μM coenzyme B 12 (*Lactobacillus* enzyme) or 11 mM Mg^{++} (*E. coli* enzyme); effectors 0.1 mM to 1 mM (ATP)

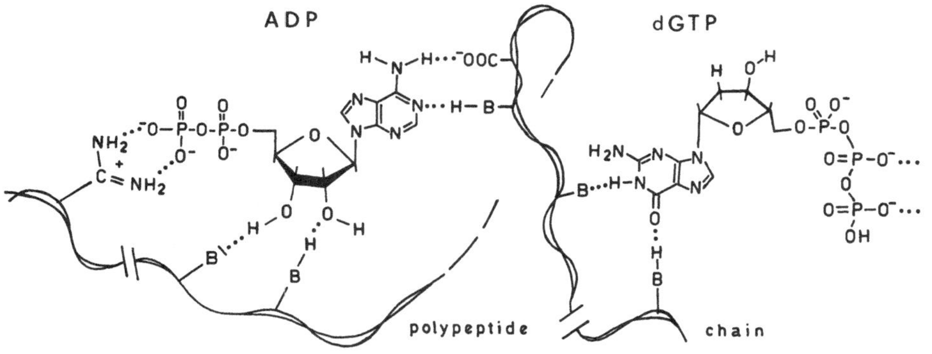

Fig. 9. A model of the interactions between a substrate ribonucleotide (ADP), an effector deoxyribonucleotide (dGTP), and ribonucleotide reductase. An underlying network of ionic forces and hydrogen bonds is proposed, as known from nucleotide – amino acid interactions in general. This model could explain the inactivity of ribonucleotides having N-methylated, *syn*-oriented, or anionic (xanthine) bases, the action of effector nucleotides through induced-fit positioning of the substrate base, and greater structural freedom observed for the effector sites[126, 135, 136]. (H)B- indicates a (protonated) basic amino acid side chain

the reduced reaction rates of methylated or unsubstituted bases. Additional hydrophobic contributions are likely for unspecific fixation of any planar base structure. Considering the size of polypeptide chains and complexity of nucleotide binding and enzyme kinetics it will be a formidable task to chemically identify the amino acids engaged in all these interactions. Such analysis, however desirable for an understanding of allosteric enzyme regulation in general, has not yet begun.

The second, hydrogen-supplying ribonucleotide reductase substrate is easier to describe in that it has to be a dithiol of sufficiently negative redox potential (Fig. 8). Monothiols such as glutathione or mercaptoethanol are inactive, as are reduced coenzymes like NADPH. All ribonucleotide reductases can function with small, unphysiologic dithiols in vitro, for example dihydrolipoate or the most commonly used dithiothreitol. The physiologic reductants are universal small polypeptides composed of about 100 amino acids, thioredoxins and, more recently discovered, glutaredoxins[9, 139] (Fig. 10). In these molecules two cysteines which are separated by two other amino acids form the dithiol. It is uncertain whether the two hydrogen transfer systems function equally well and in parallel in vivo or whether one of them is specifically programmed for ribonucleotide reduction and the other for different reductive processes. Thioredoxins and ribonucleotide reductases isolated from a wide variety of organisms are usually interchangeable and active in heterologous combinations, indicating that besides the dithiol center the rest of the molecule is of secondary importance. Like in dithiothreitol or dihydrolipoate, in *E. coli* thioredoxin the two SH groups are easily accessible as they form a protrusion at the polypeptide surface[139], ready to undergo oxidation to a 14-membered cyclic disulfide.

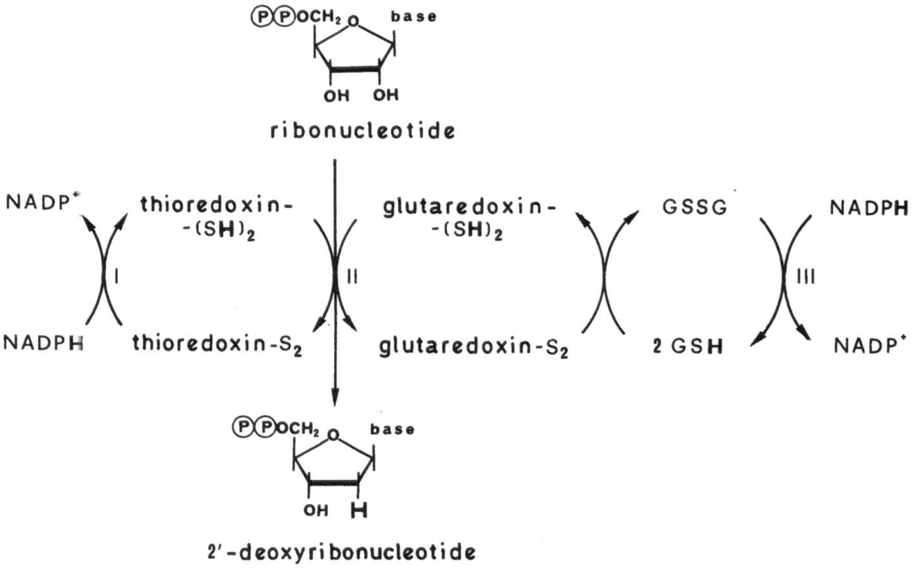

Fig. 10. Hydrogen donor systems for ribonucleotide reduction. Enzyme reactions are: I: thioredoxin reductase (EC 1.6.4.5); II: ribonucleotide reductase (EC 1.17.4); III: glutathione reductase (EC 1.6 4.2). GSH, GSSG: reduced and oxidized glutathione; NADPH, NADP: reduced and oxidized nicotinamide adenine dinucleotide phosphate coenzymes. The hydrogen transfer chain is continued in Fig. 11

2. *Nucleotides as Allosteric Effectors*

All ribonucleotide reductases, with the probable exception of a herpesvirus enzyme mentioned earlier[98)], can bind ATP and deoxyribonucleoside triphosphates in addition to their substrates, and these extra nucleotides at 0.01–1 mM concentration function as positive or negative effectors of the overall reaction. Mammalian enzymes are little active or entirely inactive (e.g., the calf thymus reductase[132)]) in the absence of activating nucleotides. The changes in enzyme kinetics are extremely complex and not even rate equations have been set up to treat such systems of two or more enzyme components, several substrates and effectors, all of different mutual affinities. However it may be stated that in general the apparent K_m value of a substrate decreases and the velocity V_{max} increases when a positive effector is present, leading to 10-fold or higher stimulation factors. Representative examples are included in Table 6. More important, these effects taken together have the consequence that slow and dissimilar reduction rates of the different substrates observed in absence of an effector approach the same magnitude. Such behaviour solves the problem of providing "equal opportunity" to all ribonucleotides with unequal base structures and is felt to be of physiologic significance for a balanced supply of DNA replication substrates. Moreover the existence of negative allosteric effectors (in particular dATP) permits feedback whenever deoxyribonucleotides accumulate. Cellular consequences of ribonucleotide reductase regulation and correlations of in vitro data with in vivo analyses of deoxyribonucleotide pools will not be

Table 6. Stimulation of ribonucleotide reductase-catalyzed reaction rates by effector nucleotides

Source of enzyme; effector added	Substrate reduction[a]				Source of enzyme; effector added	Substrate reduction[a]			
L. leichmannii[142)]	CTP	UTP	GTP	ATP	Calf thymus[132)]	CDP	UDP	GDP	ADP
none	36	14	100	20	none	0	0	0	0
dATP	102				ATP	22	14		
dCTP		37			dATP	–	–	–	–
dGTP				103	dTTP		–	18	–
					dGTP				19
T 4-phage[141)]	CDP	UDP	GDP	ADP	Human cells[133)]	CDP	UDP	GDP	ADP
none	5	3	5	3	none	11	8	4	0
ATP	30	30			ATP	185	110		
dATP	30	26			dATP	–	–	–	–
dCTP	30	18			dTTP		–	92	
dTTP			25		dGTP				110
dGTP				28					

[a] measured in % (relative rates) for *Lactobacillus* enzyme; in pmol · h^{-1} for human reductase; and in nmol · min^{-1} (V_{max}) for T 4 and calf thymus reductase. Only the action of prime effectors is indicated. The pattern of *E. coli* ribonucleotide reductase is shown in Table 7. (–) indicates inhibition

Table 7. Allosteric stimulation (+) or inhibition (–) of ribonucleotide reductase from *Escherichia coli*

Nucleotide bound to		Reduction of			
h-sites	l-sites	CDP	UDP	GDP	ADP
ATP	none	+	+		
dTTP	none	+	+	+	+
dGTP	none			+	+
ATP or dATP	ATP	+	+		
dTTP	ATP	–	–	+	
dGTP	ATP				+
any	dATP	–	–	–	–

Typical dissociation constants at 2 °C are: ATP, 10^{-5} M; dTTP, $3 \cdot 10^{-7}$ M; dATP, $0.3 \cdot 10^{-7}$ M at h-sites, $5 \cdot 10^{-7}$ M at l-sites. Data from[8, 147)]

detailed here. They are subject of excellent previous reviews and many more studies in cells and tissues since[8, 79, 143–145)].

One structural and mechanistic aspect of effector action was presented in Fig. 9, namely the concept of substrate nucleotide/protein/effector nucleotide induced fit phenomena[135)] by which optimum substrate binding and positioning for catalysis occurs. It is now seen from a comparison of the activation effects summarized in Tables 6 and 7 that there are certain preferred, stimulatory ribonucleotide/deoxyribonucleotide (or ATP) combinations. "Prime effectors"[13)] of substrate reduction are

ATP or dATP	for cytosine nucleotides (CDP, CTP)
ATP or dCTP	for uracil nucleotides (UDP, UTP)
dGTP	for adenine nucleotides (ADP, ATP)
dTTP	for guanine nucleotides (GDP, GTP)

Inhibition of substrate reduction is frequently seen in the presence of elevated dATP and dTTP concentrations. Occasionally one or the other of these effects is small or missing and additional, weaker ones are found[22, 23, 145]. Most coenzyme B 12-dependent reductases show a less complex pattern without negative regulation. However, nowhere have basically different correlations (say dGTP for CDP reduction) been observed. The same specificities hold for diphosphate and triphosphate reductases whether isolated from viral, microbial, plant, or animal sources, for B 12-dependent and iron enzymes as well as for the new manganese-activated ribonucleotide reductases[21, 22, 25, 28, 32, 44, 48, 74, 88, 132, 133, 141, 142].

It should be noted that this allosteric regulation has nothing to do with classical base pairing schemes but must be the result of twofold nucleotide/protein interaction mediated through the polypeptide chain. The above correlations are not singular but are part of the general nucleotide binding capacity of ribonucleotide reductases, demonstrated by the stimulation of reduction by dNTPs of unnatural, strongly modified substrates in the *Lactobacillus* and *E. coli* enzyme systems (Table 5).

Effector nucleotide binding to ribonucleotide reductases not only results in substrate-oriented activation as depicted in Fig. 9, but is even more complex in that other enzyme components are affected, too, and in that two unequal effector sites exist which depending on their occupation may influence each other. Such difficult-to-analyze combinations are even found in monomeric *Lactobacillus leichmannii* reductase. Here, deoxyribonucleotide binding greatly improves the enzyme's affinity for deoxyadenosylcobalamin. For example, dGTP reduces the apparent K_m value of the coenzyme from $8 \cdot 10^{-6}$ M to $3 \cdot 10^{-7}$ M in its presence[142]. All deoxyribonucleoside triphosphates are also able to catalyze the coenzyme-dependent hydrogen exchange reaction[16, 17] (Fig. 12) in the absence of substrate. Interaction with sugar-modified adenine nucleotides has shown evidence for two different binding sites[126]. O-Methyl-ATPs stimulate CTP reduction, like dATP, but do not promote hydrogen exchange while arabinonucleotides inhibit reduction[127] but do catalyze that exchange reaction. Direct measurement of binding equilibria in the absence of coenzyme, however, identified only one binding site[146].

In *Escherichia coli* ribonucleotide reductase effector nucleotides, like substrates[131], are exclusively bound to subunit B 1. Two classes of effector site were differentiated here by dialysis experiments and by chromatography on immobilized dATP or dTTP[147, 148], viz. two h (high affinity) and two l (low affinity) sites per B 1 molecule, or one h and one l site per α polypeptide chain (Fig. 2). The h-sites bind all effectors (dATP, dGTP, dTTP, and ATP) while the l-sites only interact with dATP or ATP. These findings have been integrated into a scheme where the h-sites define substrate-directed activity stimulation ("specificity sites") while occupation of the l-sites regulates the general level of enzyme activity ("activity sites") with ATP functioning as positive and dATP as negative effector (Table 7)[8, 147]. Thus if under normal metabolic conditions the l-sites are ATP-saturated, a low concentration of any deoxyribonucleotide (or again ATP) bound to the h-sites will specifically stimulate substrate reduction, as is the case when l-sites are empty. Only when dATP accumulates and is bound to an l-site it will inhibit the entire system.

Inactivation is accompanied by formation of a heavy, most likely dimeric enzyme of 15.5 S where dATP binding is tighter[147)]; this protein complex has not been characterized further.

The activation and inhibition patterns established for calf thymus and human ribonucleotide reductases resemble that of the *E. coli* enzyme in many ways (Table 6). Two different classes of effector sites are not only likely from the analogous stimulatory and inhibitory effects of deoxyribonucleotides and ATP on substrate reduction. In certain mouse cell lines two regulatory protein domains on reductase subunit M 1 have indeed been identified which are mutated independent of each other[79)]. On the other hand phage T 4 enzyme, with closely comparable protein structure, behaves different in that the inhibitory effect of dATP is missing, indicating alteration of the activity sites.

An elegant new approach to probe such binding domains was recently found in the surprisingly straightforward linkage, under UV irradiation, of [^{32}P]dTTP to specific protein residues of mouse subunit M 1[149)]. Competition experiments indicate that covalent binding occured at an allosteric site. When applied to two different mutant M 1 subunits isolated from cultured cells resistant to otherwise cytotoxic levels of deoxyadenosine or deoxyguanosine and lacking normal regulation pattern[79)], photoaffinity labeling by [^{32}P]dTTP or [^{32}P]dATP allowed to selectively block only the specificity (dNTP-binding) site or the activity (dATP, ATP) site, respectively, because the respective second site was structurally altered and unreactive in these proteins[150)]. This may permit for the first time to chemically tackle nucleotide binding site structures.

A huge amount of data, often still fragmentary, on allosteric regulation of all different types of ribonucleotide reductases awaits unambiguous analyses as the ones described. It is certainly not overstating to say that, next to the chemistry of ribonucleotide reduction, the enzymes' interaction with nucleotides and the physiologic consequences is one of the most peculiar pieces of biochemistry.

3. *Reaction Mechanisms*

The transformation of ribose to 2′-deoxyribose is reduction of a secondary alcohol to a methylene (hydrocarbon) group. If coupled with cysteine-to-cystine oxidation one calculates strongly negative free energies $\Delta G^{0\prime}$ (at pH = 7) and expects an essentially irreversible reaction. For example from the Gibbs free energies of formation $\Delta G^{0\prime}$ (25 °C), values of − 16 to − 18 kcal/mol (− 67 to − 75 kJ/mol) are estimated for reactions such as lactate (R=CH_3, R′ = COO^-) or ethylene glycol (R=H, R′=CH_2OH) reduction to propionate or ethanol, respectively (Eq. IV):

$$\underset{\displaystyle \mathrm{OH}}{\mathrm{R{-}\overset{}{C}H{-}R'}} + 2\,\mathrm{Cys{-}SH} \rightarrow \mathrm{R{-}CH_2{-}R'} + \mathrm{Cys{-}S{-}S{-}Cys} + \mathrm{H_2O} \qquad \text{(IV)}$$

However nothing is known about the activation energy for that type of reaction due to complete lack of a non-enzymatic analog. It may be substantial, considering the character of –OH as a poor leaving group and the obvious scarcity of comparable processes in carbohydrate chemistry and biochemistry.

Almost all sugar atoms of a ribonucleotide substrate (V) have been scrutinized for their participation and stereochemistry in enzyme-catalyzed reduction, and the details proved identical whether coenzyme B 12-dependent *Lactobacillus* reductase or the tyrosine-iron enzyme of *E. coli* were investigated. Reaction sequences from 2′-OH to -H passing through 2′-O-phosphorylated, ketone, 2′,3′-epoxide, and 1′,2′ or 2′,3′-unsaturated (dehydrated) intermediates could be ruled out, as reviewed previously[7].

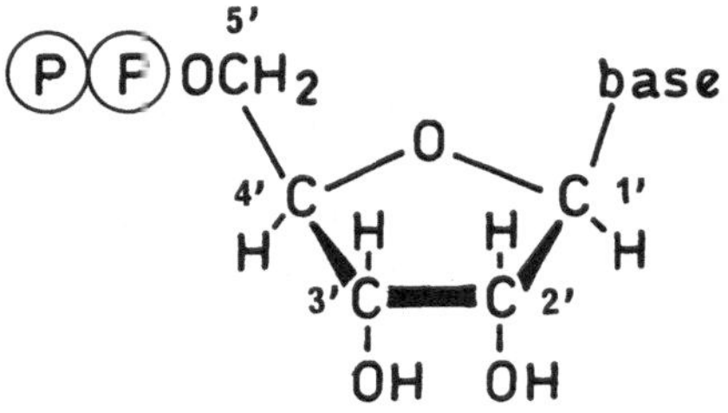

(V)

Isotope studies have always shown that only one hydrogen, derived from solvent, is introduced as 2′H of the product and that this happens with retention of configuration at C 2′[151, 152]. Follmann and Hogenkamp, using specifically labeled [^{18}O]ATPs[153] could also exclude any intramolecular oxygen migration and an S_N1 type of substitution at C 2′; the isotope in [2′-^{18}O]ATP was stable against exchange when incubated with enzyme plus coenzyme but without a reducing dithiol. Thus displacement of OH must be very direct. Stubbe has recently reconsidered the matter and advanced the hypothesis that the C 3′-H bond is reversibly cleaved during catalysis[154, 155]. She found 2- to 3-fold selection against [3′-^{3}H]UDP or -UTP when compared with unlabeled UDP or UTP, but the isotope retained its position in a product (dUTP) and little or none could be detected in water or coenzyme. The origin of these isotope effects deserves further study.

The question remains how hydrogen is transferred from a dithiol SH group to ribose C 2′-OH. (That it finally appears as derived from solvent is of course due to rapid SH/H_2O exchange.) Coenzyme B 12 or the iron/tyrosyl center, respectively, and additional amino acid residues of the enzymes are involved. Reduced thioredoxins or other dithiols do not directly donate H to a substrate but first reduce an enzyme disulfide bridge, which was demonstrated in *Lactobacillus* ribonucleotide reductase as well as in *E. coli* subunit B 1[156, 157]. The endogenous SH groups enable the enzymes to reduce a limited amount of substrate in presence of coenzyme or B 2 subunit but without another reductant. The SH groups are also target of inactivating reagents like iodoacetate, dinitrofluorobenzene, or N-ethylmaleimide[17, 156, 158] which attack cysteine residues.

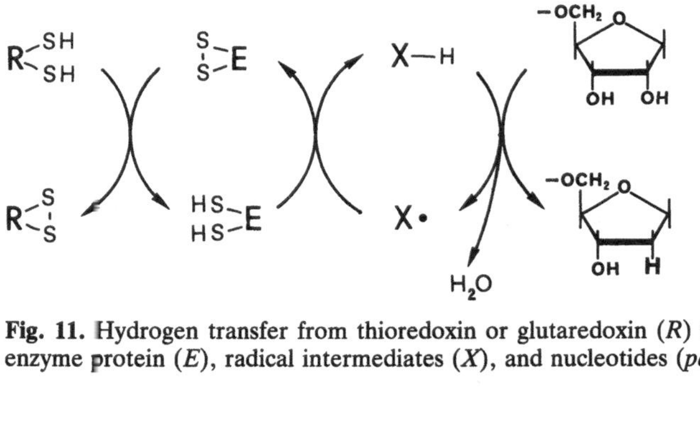

Fig. 11. Hydrogen transfer from thioredoxin or glutaredoxin (*R*) to the ribonucleotide reductase enzyme protein (*E*), radical intermediates (*X*), and nucleotides (*partial structures, right*)

Even if part of an enzyme site, an SH group cannot directly reduce OH without another activated intermediate (Fig. 11). It appears certain that "X" is coenzyme B 12 in *Lactobacillus* or the iron/tyrosyl center in *E. coli* and eukaryotic ribonucleotide reductases, respectively, and that the reactions are radical processes. The nature of oxygen-generated tyrosine radicals in the latter enzymes was described above (p. 42). We have to specify here that deoxyadenosylcobalamin (Fig. 1) can transmit hydrogens and that enzyme catalysis generates a paramagnetic radical species from the otherwise stable, low-spin cobalt(III)complex. "Hydrogen dance" sequences are a property of B 12-dependent enzymes in general[166)]. The hydrogen exchange reaction, mentioned earlier as a valuable analytical tool, and radical formation in case of *Lactobacillus* ribonucleotide reductase are summarized in Fig. 12[7, 16, 17, 160)]. In the presence of enzyme, a dithiol, and a deoxyribonucleotide effector like dGTP incorporation of hydrogen from ^3H_2O into the cobalt-bound 5'-methylene group of coenzyme occurs, or vice versa release of tritium from labeled coenzyme into water (Fig. 12, steps 1–3). Slow irreversible degradation leads to formation of 5'-deoxyadenosine and aquocobalamin (step 4), accompanied by appearance of an ESR signal of cob(II)alamin which does not parallel ribonucleotide reduction[159)]. The true intermediate of substrate reduction, however, could be observed by Blakley and others using rapid reaction techniques. Stopped-flow spectrophotometry revealed a rapid change in coenzyme absorbance at 320 and 525 nm, and that spectrum, attributed to cobalt(II) species, disappeared with $t_{1/2}$ = 145 ms after addition of substrate[160)]. With a freeze-quench method which allowed to cool reaction samples to 80 K within milliseconds it was possible to observe a new ESR signal[161)]; appearance ($k \approx 30\ s^{-1}$) and decay of the spectral change and the paramagnetic signal occured at

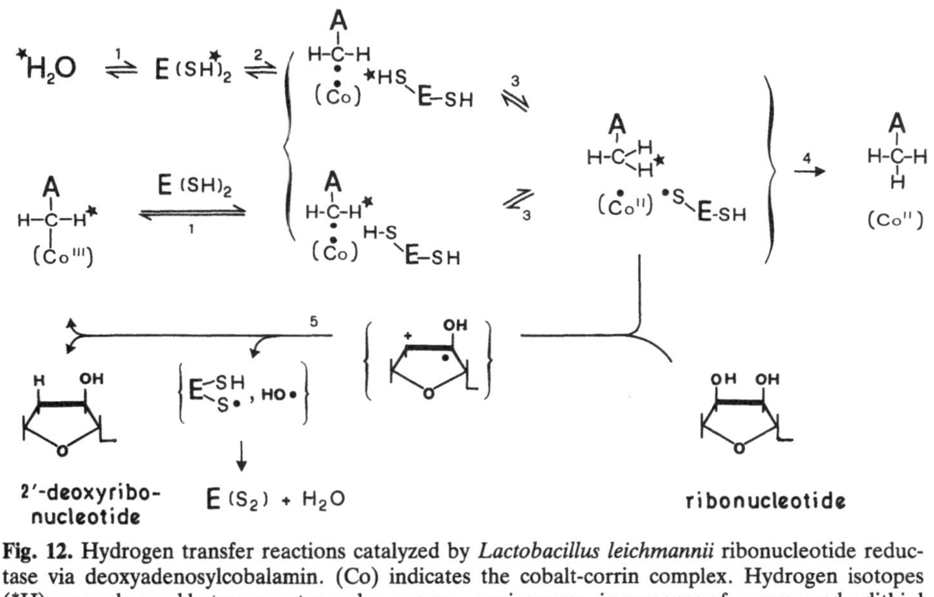

H₂O ⇌(1) E(SH)₂ ⇌(2) [A, H-C-H, (Co), *HS-E-SH] ⇌(3) [A, H-C-H*, (Co''), •S-E-SH] →(4) A, H-C-H, H, (Co'')

A, H-C-H*, (Co''') ⇌(1) E(SH)₂ [A, H-C-H*, (Co), H-S-E-SH] ⇌(3)

2'-deoxyribonucleotide ←(5) {E(SH)(S•), HO•} → E(S₂) + H₂O; {nucleotide radical} ← ribonucleotide

Fig. 12. Hydrogen transfer reactions catalyzed by *Lactobacillus leichmannii* ribonucleotide reductase via deoxyadenosylcobalamin. (Co) indicates the cobalt-corrin complex. Hydrogen isotopes (*H) are exchanged between water and coenzyme, or vice versa, in presence of enzyme and a dithiol (i.e., reduced enzyme, $E(SH)_2$) and an allosteric effector (not drawn) by reactions 1–3. Reaction 4 indicates degradation of the radical pair (center) to 5'-deoxyadenosine and cob(II)alamin. Reduction of a ribonucleotide substrate is described in step 5; the nucleotide radical (proposed in[130, 155)]) is a hypothetic intermediate

virtually the same rate. Maximum spin concentration was reached after 15–20 ms, and corresponded to about 60% of the coenzyme concentration (0.65 mM) present in the enzyme assays.

The rapidly formed ESR signal of ribonucleotide reductase-bound coenzyme B 12 is centered on a g value of 2.119 and lacks well-resolved hyperfine splitting, making it difficult to extract structural information. Extensive computer simulation[162] and the identity with the corresponding signal given by pseudocoenzyme B 12, an analog differing in the lower axial ligand of cobalt (Fig. 1) have finally led to the conclusion that "active coenzyme B 12" is the radical pair of cob(II)alamin and 5′-deoxyadenosyl radical which interact through exchange coupling and magnetic dipole-dipole coupling[163]. Homolytic cleavage of the cobalt-carbon bond (Fig. 12, 2) is well-known in the photolysis of corrinoids but it is uncertain how the dithiol-enzyme generates this system; the lower axial ligand (dimethylbenzimidazole or adenine) is probably involved because in an analog lacking the base no radical formation occurs[161]. It may also be recalled that in the threedimensional structure of coenzyme B 12 the central 5′-deoxyadenosyl substituent is surrounded by corrin methyl groups and carboxamide side chains which provide a non-polar environment facilitating radical reactions. Negative charges produced by partial hydrolysis of the carboxamides are detrimental to coenzymatic activity[7].

A logic consequence of the presence of essential radicals, transient or stable, in both ribonucleotide reductase systems is that they serve the common function to switch from a two-electron, hydride transfer chain (Figs. 10 and 11) to a one-electron, H· chain, so that the terminal transfer of H to substrate should also be a radical step (Fig. 12, 5). Abstraction of a hydrogen atom from SH (protein-bound) by the B 12 radical pair or the aryl radical of *E. coli* (all protein-bound as well) is easy to visualize, but we have no evidence for the occurence or the nature of sulfur and ribonucleotide (substrate) radicals which would ensue. Such structures are not per se unreasonable. For example a radiation-induced radical at ribose-C 5′ of cytidylic acid is known[164], and Sjöberg has recently observed a nitrogen-containing nucleotide radical during *E. coli* reductase-catalyzed decomposition of 2′-deoxy-2′-azidoCDP[130]. However the nucleotide radical cation suggested as an intermediate of the "true" reaction[130, 155] and included in Fig. 12 (step 5) is purely hypothetic, as is the nature of the leaving group. It is difficult to assess whether such nucleotide radicals, short-lived, protein-bound, and interacting with others, might be detectable by ESR spectroscopy or whether the signals, if any, were broadened beyond detection.

A reaction sequence like in Fig. 12 is also conceivable with the aryl/aryloxy radical of *E. coli* subunit B 2. (The analogy to calf thymus reductase M 2 may be even closer where a dithiol is necessary for de novo radical generation (p. 42).) In a strict sense, hydrogen exchange via radical has not been demonstrated in these enzymes, and cannot be independently measured at all because H cannot be trapped in an inert bond like $-CH_2-$ when dithiol etc. are removed. However, considering that the other components of the catalytic subunits, iron atoms, do not change valency during the reaction[37] and that the tyrosyl radical is certainly able to transmit hydrogens (Scheme III) we make the assumption that it fluctuates between phenoxyl and the hydrogenated (phenol) species for catalysis. Irreversible destruction by hydroxyurea and related compounds that function as one-electron (H·) reductants[62] is in full agreement with this view. The analogy of deoxyadenosyl and tyrosyl radicals extends to their environment where both are coordinated to metals and embedded in a fairly narrow pocket.

4. *Proteins, Nucleotides and Metal Cofactors: A Unified Model of Ribonucleotide Reduction*

Enzyme-catalyzed synthesis of deoxyribonucleotides must have preceded the appearance of DNA as the genetic material. Chemical evidence has been summarized previously[10)] that spontaneous formation of 2-deoxyribose or its 1-N-glycosides is extremely unlikely in an aqueous medium under socalled primitive earth conditions, and that the only reasonable pathway for their generation is reduction of the more readily formed, more abundant ribonucleotides. This chapter is to explore whether the properties of ribonucleotide reductases, essential catalysts in present-day biochemistry, could be reconciled with a very early origin of reductive deoxyribonucleotide formation.

Many similarities, and strong dissimilarities of the known enzymes have been mentioned above. They are all identical in that a nucleotide-binding protein, with redox-active cysteine/cystine residues, combines with another metal-containing polypeptide or a metal coenzyme in which radical intermediates can be generated and stabilized. The binding protein carries several nucleotide sites so that reaction rates are influenced by allosteric effects, with the same specificity pattern everywhere. On closer inspection it becomes apparent that the considerable individual differences in subunit composition and in the nature of the second, catalytic component can indeed be integrated into a general concept of ribonucleotide reduction.

The few reported chemical transformations of a sugar, ribonucleoside or -nucleotide to 2-deoxysugar or 2′-deoxyribonucleos(t)ide, usually carried out with highly substituted intermediates in organic solvents, have in common that sulfur derivatives (thioethers or thioesters) and metal-catalyzed (Ni, Pd, $(C_4H_9)_3SnH$) hydrogenation steps are best suited to permit deoxygenation of a C–OH group. Moreover N-glycosides are more favorable than sugars, probably because the heterocyclic bases stabilize intermediates or transition states and direct the reaction to C2′. In the most efficient procedure so far, achieved by Robins[165)], ribonucleoside-2′-phenoxythiocarbonates were reduced smoothly to 2′-deoxyribonucleosides by treatment with tri-n-butyltin hydride (tributylstannane) and a radical starter in toluene at 75 °C. Although not direct models for biochemical conditions, the analogy of such reactions to enzymatic ribonucleotide reduction with respect to sulfur intermediates and radical-(metal-)induced cleavage of the C2′-O bond appears more than superficial.

The most peculiar property of the whole group of ribonucleotide reductases is their dependence on transition metals, either manganese, iron, or cobalt (cobalamin) for catalytic activity. The mode of action of Mn^{++} in a bacterial reductase subunit is not yet known; however in view of the exchangeability with Fe^{++} (or, protein-bond, Fe^{+++}) in two enzyme systems in vitro[44, 56)] and their similar complex chemistry including Me-O-Me complexes we feel that these two isoelectronic metal ions (Mn^{++}, Fe^{+++}) should have closely comparable function even if differentially coordinated in the native enzymes. The iron-containing B2, M2 etc. subunits of bacterial and eukaryotic reductases, and deoxy-adenosylcobalamin as the coenzyme of other microbial enzymes are of course entirely different in structure. However it cannot be overlooked that they serve the same catalytic function to provide a hydrogen-transferring radical, either tyrosyl residue coordinated to the binuclear iron complex, or deoxyadenosyl residue coordinated to the cobalt-corrin complex. The metals are thought to facilitate radical formation through changes in their ligand fields. Conformation changes ("on"/"off") of the lower axial benzimidazole or adenine ligand possibly accompany the coenzymatic function of deoxyadenosylcobalamin[163]. The geometries of iron complex and tyrosine in the other ribonucleotide reductases are unknown. However there is good evidence from ESR work[167] that formation, life time, and H transfer reactions of free radicals, including phenoxy radicals generated from unsubstituted phenols and molecular oxygen, are greatly favoured in the coordination field of transition metal complexes.

We propose that virtually all ribonucleotide reductases are built and function in the same way: They consist of a nucleotide-binding SH enzyme and a cofactor which may be a polypeptide or a big, complex coenzyme. There are several reasons besides the functional aspect discussed above which indicate that the catalytic iron proteins are *not* classical enzyme subunits but may be viewed as cofactors. In eukaryotes they are present in nonstoichiometric amount, and are synthesized independent of the larger enzyme subunit[59, 71]. Most catalytic subunits dissociate readily from holoenzyme during column chromatography even under mild conditions. Of the two *E. coli* subunits, B 2 has a high (~ 60%) content of helical polypeptide, typical of non-enzyme proteins such as hemoglobin or hemerythrin, whereas the low helix percentage of B 1 is that of a typical enzyme[39]. In kinetic studies with *Lactobacillus* and *E. coli* ribonucleotide reductase we have seen a remarkable similarity in the dependence of reduction rates on coenzyme or subunit B 2 concentration, respectively (Fig. 13): In the absence of an effector nucleotide but with saturating substrate, dithiol, and Mg^{++} this subunit seems only poorly bound to B 1. (In fact the 1 : 1 complex of B 1 and B 2 was only demonstrated in presence of dTTP[33].) Binding of an effector nucleotide to B 1 obviously increases its affinity for B 2 just like an effector increases the affinity (decreases K_m) of deoxyadenosylcobalamin and *Lactobacillus* apoenzyme. Vice versa the presence of B 2 had no effect on the binding equilibria of nucleotides with *E. coli* protein B 1[147]. All these observations support the idea that the larger, nucleotide-binding and SH-containing proteins (where in fact both substrate and reductant are assembled) interact with the radical/metal polypeptides in an apoenzyme-coenzyme-like fashion.

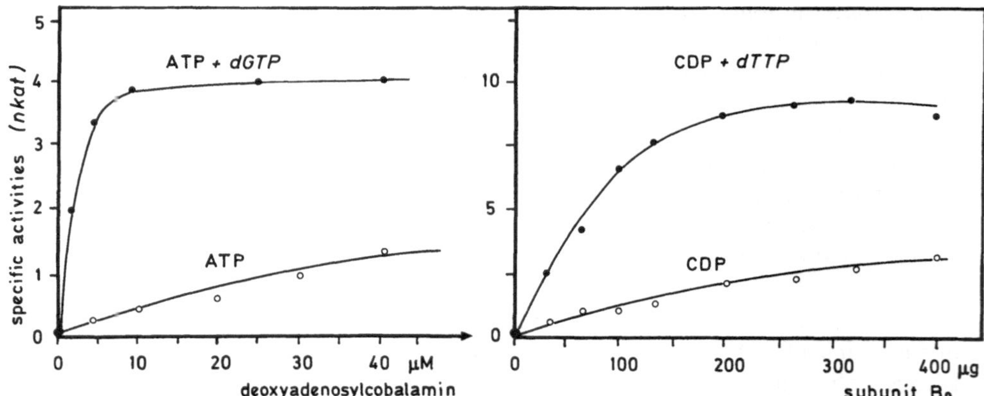

Fig. 13. Dependence of the specific activities of *Lactobacillus leichmannii* (left) and *Escherichia coli* (right) ribonucleotide reductases upon coenzyme B 12 and subunit B 2 concentration, respectively, in the absence and presence of allosteric effector nucleotides (dGTP for ATP reduction, dTTP for CDP reduction). Conditions as in Table 5

Inspection of the size of the various nucleotide-binding proteins, whether coenzyme B 12- or iron subunit-associated, reveals a preference for the 75,000 to 100,000 dalton range. The 45,000 dalton L 1 subunits of rat liver reductase with one half this value still fit the picture although in no other case dissociation into two polypeptide chains has been found. One, two, or four of these large subunits can apparently form a holoenzyme like in the *Lactobacillus* (1), *Corynebacterium, E. coli, Scenedesmus,* mammalian enzymes (2) or *Rhizobium* ribonucleotide reductase (4 subunits). Such large size of one polypeptide (comparable, for example, to that of DNA polymerase I) ist obviously required for several nucleotide sites, thioredoxin, and coenzyme B 12 or iron subunit binding.

It is less clear why the cofactor-like subunits of iron enzymes also have to be large proteins. Polypeptide chains of 35,000 to 55,000 dalton predominate here. They usually occur in pairs of two but there are cases where only one subunit of M_r 75,000 to 80,000 is found[48, 68] which is not further dissociable. The size could be favorable for integration of the reductase holoenzyme into even larger multienzyme aggregates, discussed in Sect. E.

With this knowledge of the components of ribonucleotide reduction it is not too difficult to imagine how such an enzyme apparatus could have assembled under prebiotic conditions. Nucleotide binding domains in polypeptides are a property of many ancient enzymes, as are short chain segments with two cysteines or a cystine loop, respectively. Both structures combined in a primitive protein and interaction with a corrinoid (assumed to have existed as early as or earlier than porphyrins[117]) would constitute a reductase of the (anaerobic) *Lactobacillus*, or cyanobacterial type. Alternatively various types of metal-binding polypeptides are also a primitive class of compounds, and combination of a transition metal cluster with tyrosine residues and catalytic amounts of oxygen can have led to the catalytic subunit type. Although fairly unique in biochemistry, these proteins may not be too complex in structure as judged by the simple inactivation and reactivation conditions observed with eukaryotic reductases (but not the possibly more specialized, more recent *E. coli* enzyme). Thus the nucleotide-binding SH protein could also have combined with iron, or manganese containing catalytic subunits. Consequently ribonucleotide reductases appear of polyphyletic origin with respect to their catalytic cofactors, and it would not be too big a surprise if yet another one was found in the microorganisms still unstudied.

The essentials of an ur-reductase are put together in Fig. 14. It may have been even simpler if allosteric effectors were not included.

The idea of common ancestry of the different ribonucleotide reductases is difficult to test at present because protein sequencing studies have not yet begun. Exchangeability of subunits, possible among the calf thymus and mouse enzymes[60], has not much been tried. We have seen small but significant stimulation of enzyme activity when the separated, inactive subunits U 1 of algal (*Scenedesmus*) ribonucleotide reductase and B 2 of *E. coli* were recombined, but not in the reverse combination (B 1 + U 2)[48]. The many individual differences in enzyme structure like Mg^{++} or Ca^{++} requirement for subunit interaction, variations in the radical environment as expressed in slightly different ESR spectra (Fig. 3), or details of allosteric effector pattern, do not in principle contradict our reductase model but will in reality severely limit its experimental verification.

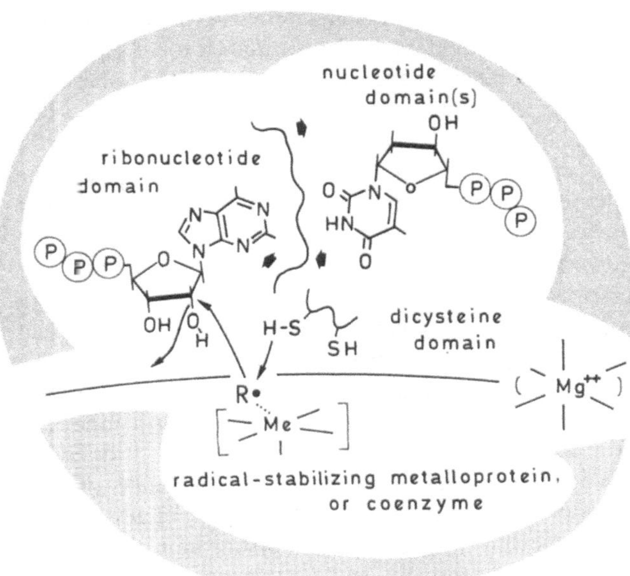

Fig. 14 Structure and functional domains of an "ur-reductase". Such an enzyme is thought to combine two or three nucleotide binding sites (which affect each other) with a dicysteine center in one polypeptide chain, and a metal-coordinated radical in another polypeptide, or coenzyme; interaction of the two components may, but must not involve Mg^{++} or other ions

D. Inhibitors of Ribonucleotide Reduction

1. Targets of Enzyme Inhibition

The complex structure and molecular mechanism of ribonucleotide reductases make them vulnerable to many kinds of distortion. Not only has it been an "... experience that T4 ribonucleotide reductase cannot stand a 0.5 M concentration of anything except water"[91] but enzymes from all sources are easily inhibited by one or several of the following interferences:

a) blocking of substrate or effector nucleotide binding sites,
b) inactivation of the catalytically active dicysteine system,
c) withdrawal of metal ions with structural or catalytic function,
d) destruction of, or hindered generation of the radical,
e) disturbance of native protein structure at other domains important for activity.

The first category comprises "suicide substrates" (2-deoxy 2′-azidoCDP etc.), or effector analogs like arabinonucleotides, and the second category SH reagents, both mentioned above (p. 50, 58). Inhibition by c) and d) is often structurally and mechanistically interdependent, and an inhibitor may be both a chelator and a radical scavenger. The last group comprises all the less specific, denaturing effects of ionic strength, divalent metal ions or other additions which lead to dissociation or aggregation of subunits or otherwise unfavorable protein conformations. It is these subtle, frequently unknown parameters that make optimum conditions mandatory for the intricate enzyme assays.

Besides inhibition of more general kind (a, b, e) no selective inhibitors of the coenzyme B 12-dependent ribonucleotide reductases have become known. Most modified analogs of deoxyadenosylcobalamin are inactive, few are inhibitory; their action has been reviewed[7]. The following discussion therefore concentrates on the iron/radical type of ribonucleotide reductase. One should assume metal chelators to be important inhibitors of these enzymes but this anticipation is not fulfilled. EDTA, the standard chelating agent in biochemical research acts only at high concentrations or in some cases even stimulates the reaction. Iron-specific chelators are about an order or magnitude more effective, however several enzymes prove to be insensitive or resume activity as soon as the chelator is removed. Inhibition of the manganese-requiring reductase of *Brevibacterium ammoniagenes* by cyanide (Fig. 5) appears to represent the first example of truly metal-directed enzyme inhibition. Table 10 summarizes present knowledge of ribonucleotide reductase inhibitors.

2. *Hydroxyurea and Analogs*

Hydroxyurea blocks DNA synthesis in many organisms with only minor impact on RNA and protein synthesis. It was the first synthetic inhibitor of ribonucleotide reduction to be recognized[169–172] and has since been frequently applied in mechanistic studies, in cell biology, and, to a lesser extent, in clinical medicine as an inhibitor of DNA replication and cell proliferation. The cellular aspects have been reviewed by Timson[173]. Our present discussion centers on the molecular mechanism of action, and on the important question whether the many diverse effects observed in vivo can be traced back to one single cause; this matter is afflicted with some confusion.

Hydroxyurea, and hydroxamates in general are simple yet reactive chemicals (Fig. 15). The hydroxamate moiety –CO–NOH– is capable of chelating metal ions with particular affinity to ferric iron[174–177] whereas the hydroxylamine moiety >N–OH under-

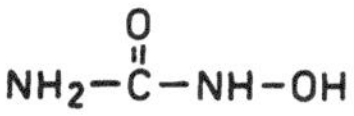

$NH_2-\overset{\overset{O}{\|}}{C}-NH-OH$

I: hydroxyurea

$R-\overset{\overset{O}{\|}}{C}-NH-OH$

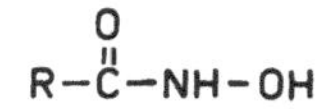

II: hydroxamic acids
R = alkyl or aryl

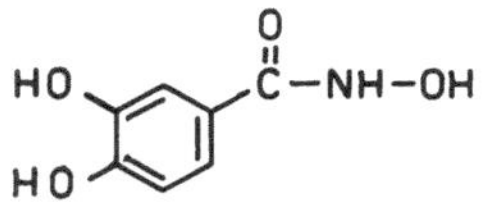

$(HO)_2C_6H_3-\overset{\overset{O}{\|}}{C}-NH-OH$

III: 3,4-dihydroxybenzohydroxamic acid

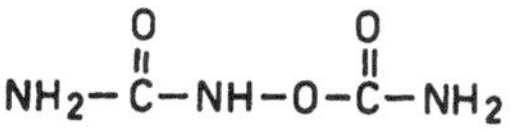

$NH_2-\overset{\overset{O}{\|}}{C}-NH-O-\overset{\overset{O}{\|}}{C}-NH_2$

IV: carbamoyloxyurea

$R-\overset{\overset{COOH}{|}}{CH}-NH-OH$

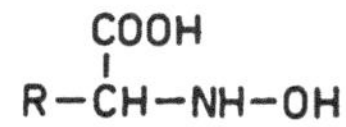

V: N-hydroxy-α-amino acids
R = alkyl

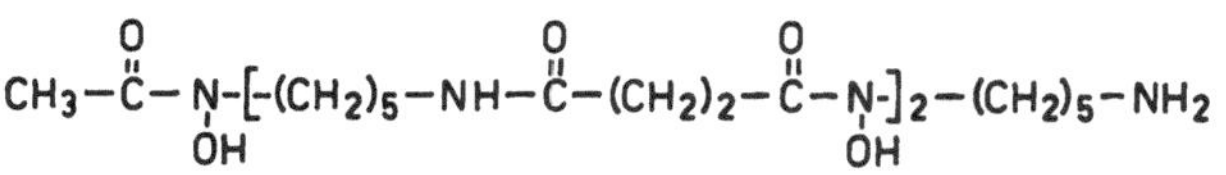

$CH_3-\overset{\overset{O}{\|}}{C}-\underset{\underset{OH}{|}}{N}-[-(CH_2)_5-NH-\overset{\overset{O}{\|}}{C}-(CH_2)_2-\overset{\overset{O}{\|}}{C}-\underset{\underset{OH}{|}}{N}-]_2-(CH_2)_5-NH_2$

VI: desferrioxamine B (mesylate salt: Desferal®)

Fig. 15. Structure of hydroxyurea and related compounds

goes one-electron oxidation involving free radicals[62, 178–180]. All the compounds listed in Fig. 15 have been found to inhibit ribonucleotide reduction, viz. aliphatic and aromatic hydroxamic acids[62, 181], N-hydroxy-α-amino acids[48], N-carbamoyloxyurea[182] and desferrioxamine[183, 184]. The latter tris-hydroxamate of microbial origin is an extraordinarily potent iron chelator[185]. However the most efficacious inhibitors of this class are the polyhydroxybenzohydroxamic acids (e.g. III in Fig. 15) with a 50% inhibition concentration, $I_{50} = 10^{-5}$ M for mammalian enzymes[62, 181], making them candidates for chemotherapy.

The effects of hydroxyurea in the purified ribonucleotide reductase systems of *E. coli*, phage T4, calf thymus and mouse cells have been described above (p. 36, 42). Inhibition of substrate reduction in vitro ($I_{50} = 2 - 3 \cdot 10^{-4}$) is accompanied by loss of the tyrosyl radical, but not iron from *E. coli* subunit B2. Studies with substituted hydroxylamines and hydroxamates showed good correlation between their ability to undergo one-electron oxidation and enzyme inhibition, unless branched substituents prevented interaction with the protein (Table 8)[62]. Thus the mode of inhibition of *E. coli* ribonucleotide reductase is essentially solved: Within steric restrictions of accessibility to the active site the compounds donate an electron to the enzyme's free radical, producing an inactive protein with still intact binuclear iron complex (Eq. VI). This process is irreversible in vitro until iron is removed, and then reintroduced with Fe(II)ascorbate in the presence of oxygen, whereupon radical and enzyme activity reappear. No other enzyme of *E. coli* has been found to be inactivated by hydroxyurea.

$$\text{—C}_6\text{H}_4^{+\bullet}\text{—OH or —C}_6\text{H}_4\text{—O}^\bullet \ [\text{Me}] + \text{RR'N—OH} \rightarrow \text{—C}_6\text{H}_4\text{—OH} \ [\text{Me}] + \text{RR'N}\dot{-}\text{O} \ (+ \text{H}^+) \quad \text{(VI)}$$

Table 8. Correlation of one-electron oxidation rate[a] and inhibition of *E. coli* ribonucleotide reductase by hydroxyurea and hydroxamates. Data from [62]

Compound R-NHOH	R	$t_{1/2}$ (s)	% enzyme inhibition
urea (control)		(> 1 h)	0
hydroxyurea	NH_2CO	< 5	99
N-methylhydroxylamine	CH_3	< 5	94
acetohydroxamic acid	CH_3CO	25	82
n-hexanohydroxamic acid	n-$C_5H_{11}CO$	41	64
benzohydroxamic acid	C_6H_5CO	29	32
isobutyrohydroxamic acid	iso-C_3H_7CO	27	10
cyclohexylhydroxamic acid	$C_6H_{11}CO$	40	0

[a] measured spectrophotometrically by reaction with the free radical salt, potassium nitrosodisulfonate

On the cellular level, treatment of *E. coli* cultures with the inhibitor leads to immediate depletion of the small deoxyribonucleotide pools without gross alterations of ribonucleotide concentrations[186–188]. DNA replication and cell division cease in parallel at $2 - 3 \times 10^{-2}$ M hydroxyurea in the growth medium[186, 188, 189], which for unknown reasons is about 10 times the concentration needed for total inactivation of ribonucleotide reductase in vitro[91]. Unlike the latter case, in vivo inhibition is readily reversed upon removal of the drug[188, 189], even if protein synthesis is blocked[190], demonstrating that the cells can reactivate their enzyme molecules. Deoxyribonucleotide pools and DNA synthesis cannot be restored in the presence of hydroxyurea by simultaneous addition of deoxyribonucleosides[189], probably because *E. coli* lacks appropriate kinases for salvage[191]. Inhibition is overcome, however, if permeabilized cells are supplied with deoxyribonucleotides[188]. This finding does not exclude interference of hydroxyurea with other metabolic processes but proves DNA polymerization to be insensitive. Since inhibitor concentrations sufficient to suppress DNA synthesis and cell proliferation only slightly affect overall RNA and protein synthesis[188, 189, 192], cell mass continues to increase, producing abnormally elongated cells[189] which after a prolonged state of "unbalanced growth" lose viability[188]. This little understood lethal damage in response to hydroxyurea requires protein but not RNA synthesis[192], can be mutationally decreased[193], and possibly includes alterations of DNA structure[194]. Best evidence that it is indeed ribonucleotide reductase which determines sensitivity of *E. coli* towards hydroxyurea comes from genetic studies. Cell proliferation of mutant strains defective either in the iron/radical subunit B 2 or in both subunits is decreased at lower drug concentrations than needed for control strains[31, 195, 196] while cells harbouring recombinant plasmids with functional genes of both subunits which overproduce the enzyme acquire resistance[31]. In conclusion, the inhibition of both DNA synthesis and cell proliferation in *E. coli* by hydroxyurea can be sufficiently explained by the inactivation of ribonucleotide reductase observed in vitro.

DNA synthesis and growth of many other genera of prokaryotes are suppressed by hydroxyurea[173, 189], and sensitivity is frequently taken as evidence for the presence of an *E. coli*-like ribonucleotide reductase. This conclusion has now to be extended to manganese-requiring ribonucleotide reductases of *Brevibacterium ammoniagenes*[44] and the other bacterial species listed in Table 4. Even organisms with a deoxyadenosylcobalamin-dependent ribonucleotide reduction may be affected by hydroxyurea despite the insensitivity of this class of enzymes in vitro[22, 171]. *Bacillus megaterium* has been claimed to be sensitive (cited in[197]), and greening and cell division of *Euglena gracilis* are inhibited by the drug[121, 122]. Obviously the inhibitor is able to interfere with cellular processes other than ribonucleotide reduction and is not as specific as commonly thought; hence it is of limited use for identifying a particular type of ribonucleotide reductases.

All eukaryotic ribonucleotide reductases assayed so far in vitro are inhibited by hydroxyurea, and a wide variety of eukaryotic cells and organisms is found susceptible to DNA synthesis inhibition by the drug[173]. Nevertheless, a clearcut dependency like in *E. coli* cannot be stated as investigation of the eukaryotic enzymes is lagging and because hydroxyurea exerts manifold cellular effects, some without obvious relation to the inhibition of ribonucleotide reduction (cf. Table 9).

The in vitro reduction of CDP, ADP, and GDP by different mammalian enzymes is as sensitive to hydroxyurea ($I_{50} = 2 - 4 \times 10^{-4}$ M) as with *E. coli* enzyme[62, 76, 88, 171, 172, 182]. The tyrosine radical of mouse fibroblast ribonucleotide reductase disappears upon addition of the inhibitor in vitro as well as in vivo indicating an analogous mode of action as with the bacterial enzyme[59,60]. Comparison of the inhibitory potency of a series of polyhydroxybenzohydroxamic acids towards rat tumor ribonucleotide reductase showed the most effective radical scavengers to be the best inhibitors[198], and radical destruction was directly established with the most potent substance, 3,4-dihydroxybenzohydroxamic acid (Fig. 15, III)[62]. The ease of reactivation differs substantially among individual ribonucleotide reductases. The enzyme of *E. coli* is reactivated under special conditions[37], maybe as a consequence of its relatively inaccessible radical center[62]. Inhibition of eukaryotic enzymes is more readily reversed when hydroxyurea is removed and a dithiol, oxygen (or air, respectively), and iron ions are present, i.e. under standard assay conditions. Requirements for added ferrous, or ferric iron appear variable, ranging from undetectable in the yeast reductase system[52] to stimulation in mouse cell enzyme[60] and an indispensable role in calf thymus[346] and algal ribonucleotide reductase (Fig. 7)[48] where the metal is apparently lost from the enzyme.

The latter example demonstrates that the metal-chelating ability of hydroxyurea, although inferior in significance to its role as radical scavenger, cannot be neglected. It would be interesting to know how the drug or its analogs affect other radical or metal targets in cellular metabolism besides ribonucleotide reduction, but here the experimental evidence is scanty. Nickel-containing urease is inhibited by hydroxyurea and hydroxamates[199,200,349], and zinc-containing proteases are inactivated by amino acid-derived hydroxamates[201,202]. Aromatic hydroxamates inhibit metal-dependent redox enzymes such as peroxidase and tyrosinase[203], soy bean lipoxygenase[204], and others[205,206] by as yet unknown mechanisms. In contrast, the following enzyme systems involved in eukaryotic DNA metabolism, metal-containing or not, were not affected in vitro by hydroxyurea concentrations that effectively inhibit DNA synthesis in vivo: thymidylate synthase[207–209], nucleotide kinases[210,211], DNA polymerases[210–212], DNA ligase[213], DNA methylase[208,214,215], or ornithine decarboxylase[216].

Should the very simple reduction of a tyrosyl radical in ribonucleotide reductases (Eq. VI) really be the only interception of hydroxyurea with a cell's replication machinery, it has to stand up against an impressive list of consequences in eukaryotic cells. A selection of such effects, mostly observed in cell cultures, is compiled in Table 9.

We will only comment on the more prominent features of all this work with a special view on their relationship to ribonucleotide reduction. Inhibition of mitotic and, at higher drug concentration, of repair DNA synthesis is expected and shows them to depend upon the same source of precursors. The sensitivity of semiconservative and repair synthesis differs by about an order of magnitude, and this permits use of hydroxyurea in studies of DNA repair to suppress interference from replication. In most mammalian cells inhibitor concentrations of about 2×10^{-3} M prevent DNA replication, cell proliferation, and in vitro ribonucleotide reduction equally well. In several cases inhibition can be overcome by concomitant supply of deoxyribonucleosides[173,226,253,265]. Failure of reversal in other systems may be due to inappropriate use of exogeneous purine deoxyribonucleosides[211,253,267] or different metabolic channeling of salvage and de-novo nucleotides[255]. As hydroxyurea rather selectively inhibits DNA synthesis, only cells in S-phase are arrested while those in other compartments of the cell cycle continue to grow until they accumulate early in S-phase. Similarily, cell death is confined to S-phase. As long as cells have not been lethally damaged, inhibition is immediately reversed upon removal of the drug, a property often utilized in cell synchronization schemes. Suffi-

Table 9. Effects of Hydroxyurea on Eukaryotic Cells

	Selected references
a) inhibition of mitotic DNA synthesis without severe effects on RNA and protein synthesis ("unbalanced growth")	173, and many practical applications
b) action on permeabilized cells	83, 217, 218
c) inhibition of premeiotic DNA synthesis	219–221
d) interference with DNA repair after UV lesions	173, 213, 218, 222–225
e) altered cell morphology (consequence of unbalanced growth)	211, 226–232
f) cell cycle specific action, cell cycle arrest, cell synchronisation	173, 229, 232–238, and practical applications
g) inhibition of histone synthesis	230, 239–242
altered activity of cell cycle-coordinated enzymes:	
ribonucleotide reductase	82, 198, 244
thymidine kinase	184, 245, 246
ornithine decarboxylase	216
h) selection of mutants with elevated or altered ribonucleotide reductase	59, 60, 247 (review), 248
i) alteration of deoxyribonucleotide pools	249–256
j) changes in nucleotide metabolism:	
decreased in situ thymidylate synthase activity	207, 208
altered salvage reactions	184, 245, 246, 253, 257, 258
k) residual synthesis of low molecular weight DNA	221, 222, 233, 259–263, 350
l) accumulation of single-stranded DNA	264
DNA strand breaks	259, 265
altered chromatin and chromosome structure	173, 227, 230
m) S-phase specific cell death	173, 235, 245, 263
n) teratogenesis and mutagenesis	124, 173, 266

ciently low concentrations of hydroxyurea do not block cell cycle progress but slow down DNA replication, thus extending S-phase at the expense of G_1[232, 238]. The unbalanced growth during inhibition produces morphological changes like increase in cell volume and nuclear swelling.

In HeLa cells hydroxyurea is an efficient inhibitor of histone synthesis. This action requires protein synthesis and leads to rapid disappearance of cytoplasmatic histone mRNA[239]. The effect is not specific for hydroxyurea since suppression of DNA synthesis by arabino-cytosine[268] or temperature-sensitive mutations[269, 270] yields analogous results. Similarly, the synthesis of some enzymes necessary for DNA replication and active in S-phase is altered by hydroxyurea. Increased activity of ribonucleotide reductase in HeLa and in hamster cells[198, 244] and of the salvage enzyme thymidine kinase in HeLa cells and KB cells[245, 246] has been observed, probably as a consequence of the increased fraction of cells in S-phase. Repression occurs for thymidine kinase in human lymphocytes[184] and for ornithine decarboxylase in Chinese hamster fibroblasts[216] whereas no or only slight effects were seen on ribonucleotide reductase in hamster fibroblasts[82], on thymidylate synthase in extracts from synchronous mouse cells[207, 243], and on DNA polymerase in rabbit kidney cells[271] or HeLa cells[245, 272].

Changes in cellular deoxyribonucleotide pools following addition of hydroxyurea are the most probable direct cause for the halt of DNA synthesis. However, as the pools are differentially affected with the purine deoxyribonucleotides usually decreasing and the pyrimidine deoxyribonucleotides somewhat increasing, despite identical in vitro sensitivity of reduction of all substrates, this fact has cast doubt on the intracellular mode of action of hydroxyurea. The apparent contradiction may be resolved by the assumption that residual DNA synthesis in the presence of inhibitor exhausts the smallest precursor pool[233] which is often dGTP or dATP, while the remaining ribonucleotide reductase activity leads to accumulation of the pyrimidine deoxyribonucleotides in spite of decreased overall synthesis. This may be favoured by depletion of the negative allosteric effectors dATP and dGTP. Different intracellular stabilities of the DNA precursors are conceivable, too. Recent experiments suggest that hydroxyurea not only inactivates ribonucleotide reductase but also affects the organisation of the multienzyme aggregate of precursor and DNA synthesis[83, 273] (cf. Sect. E), leading to further changes in nucleotide metabolism like decreased in vivo activity of thymidylate synthase.

Even during exposure of cells to higher concentrations of hydroxyurea some DNA is synthesized but accumulates as replication intermediates (of Okazaki fragment or up to replicon size) which upon removal of the drug can be chased into bulk DNA. How does the inhibitor bring about this puzzling effect? There is no evidence for direct action on the ligating enzymes[213], however it is conceivable that in an altered state of the replication complex of DNA polymerase and precursor synthesizing enzymes the polymerisation and ligation steps are uncoupled. The depleted substrate pools may then permit residual DNA synthesis but are not available for the ligation of DNA pieces. The partially replicated state of DNA may be especially vulnerable to nuclease attack and its prolonged persistence leads to lethal disaggregation of chromatin, thus ultimately linking ribonucleotide reductase inhibition to cell death[263].

3. *Aromatic and Heterocyclic Compounds*

As ribonucleotide reduction is the indispensible source of DNA precursors its inhibition is an appealing goal for antiproliferative chemotherapy. Mainly for this reason numerous chemicals are constantly explored as inhibitors of the enzyme, in particular in the laboratories of Cory, Elford, and Moore. Unfortunately, although some highly efficient compounds were discovered none of them has proved useful in clinical application. These findings, together with those from more biochemical approaches are compiled in Table 10.

The most obvious common feature of all ribonucleotide reductase inhibitors except for the least effective chelators desferrioxamine, EDTA, and EGTA is that all are aromatic or heterocyclic. The superior activity of aromatic compounds became evident in comparative studies with hydroxamate derivatives, too[62, 181, 198]. A corresponding hydrophobic binding site on mammalian ribonucleotide reductases was postulated a decade ago by Sartorelli and Moore in their investigation of substituted α-(N)-heterocyclic carboxaldehyde thiosemicarbazones[293]. If it exists ist must be located on the iron/radical protein (M2 etc.) as several of the inhibitors selectively act on this subunit[62, 281, 284, 286]. The lack of competition between guanazole or the thiosemicarbazones and nucleotides

Table 10. Inhibitors of ribonucleotide reduction (cf. Fig. 15 for hydroxyurea and analogs)

Type of compound	Inhibits enzyme of	I_{50} (M)[a]	Mode of action, where known	Ref.
a) α-(N)-heterocyclic carboxaldehyde thiosemicarbazones, resp. their ferrous complexes				
pyridine-2-carboxaldehyde thiosemicarbazone	Novikoff hepatoma of rat	4×10^{-7}	reversible, competes with dithiol	274
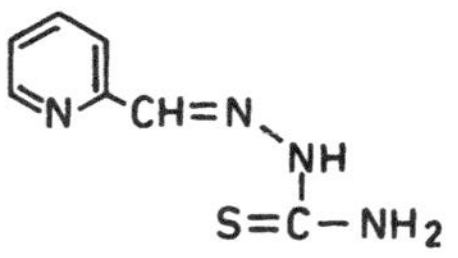	same	9×10^{-7}		275
	Sarcoma 180 cells of mouse	6×10^{-7}		274
	H. Ep. 2 human carcinoma cells	3×10^{-7}		276
	same			277
	E. coli	$>10^{-3}$		274
isoquinoline-1-carboxaldehyde thiosemicarbazone	Novikoff hepatoma of rat	5×10^{-8}	partially reversible, competes with dithiol	278, 279
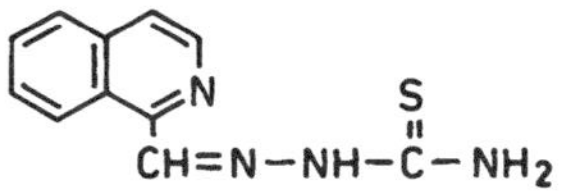	same	4×10^{-7}		275
	Sarcoma 180 cells of mouse	7×10^{-8}		278
	Ehrlich ascites tumor of mouse	3×10^{-7}		280
	H. Ep. 2 human carcinoma cells	2×10^{-7}		276
or	human tumor cells	4×10^{-7}		87
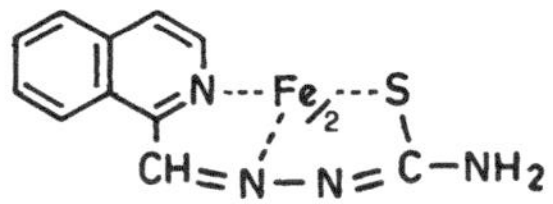	calf thymus	(8×10^{-8})		281
	mouse fibroblast cells		oxygen-dependent destruction of free radical	281
	inactive with *E. coli* enzyme	–	–	278
4-methyl-5-amino isoquinoline-1-carboxaldehyde thiosemicarbazone	Novikoff hepatoma of rat	6×10^{-8}	reversible without metal addition	282, 283
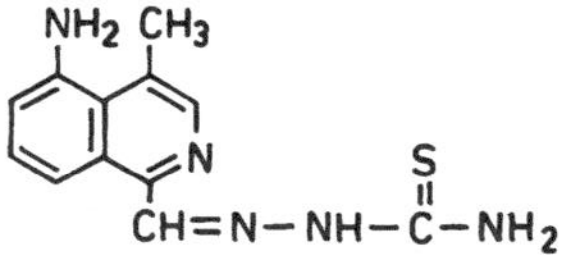	same	6×10^{-7}		275
	Ehrlich ascites tumor of mouse		acts on catalytic (iron) subunit	284
	same	$\approx 10^{-7}$	partially reversible without metal addition	280
	E. coli	$>2 \times 10^{-5}$		283

Table 10. (continued)

Type of compound	Inhibits enzyme of	I_{50} (M)[a]	Mode of action, where known	Ref.
b) phenols				
pyrogallol derivatives R=H, $CONH_2$, $COOCH_3$	Novikoff hepatoma of rat	$\approx 10^{-5}$	radical scavenger?	181, 198
catechol derivatives R=CH_2NH_2, $CONH_2$, $COOCH_3$	Novikoff hepatoma of rat	$\approx 5 \times 10^{-5}$		181, 198
3-methyl-4-nitrophenol	Chinese hamster ovary cells	2×10^{-3}		285
aurintricarboxylic acid	*Scenedesmus obliquus* *Saccharomyces cerevisiae* *E. coli*	2×10^{-6} 3×10^{-5} 2×10^{-6}		48 52 48

c) miscellaneous aromatic compounds

guanazole (3,5-diamino-1H-1,2,4-triazole)	H. Ep. 2-human tumor cells	5×10^{-4}	analogous to hydroxyurea	183
	Chinese hamster ovary cells	6×10^{-4}	analogous to hydroxyurea	182
	mouse L cells	5×10^{-4}	analogous to hydroxyurea	76
	calf thymus	5×10^{-4}	analogous to hydroxyurea	56, 62
	mouse fibroblast cells		destruction of free radical	62
2,3-dihydro-1H-pyrazolo[2,3-a]imidazole 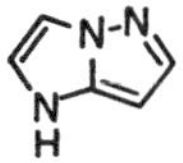	Ehrlich ascites tumor of mouse	$\approx 4 \times 10^{-4}$	acts on catalytic subunit, reactivation by Fe(II)	286
2-nitro-imidazole (azomycin) 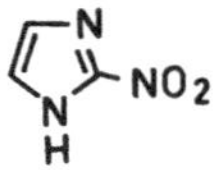	*E. coli*	$\approx 5 \times 10^{-4}$		287
4-amino-2-phenylimidazole-5-carboxamide 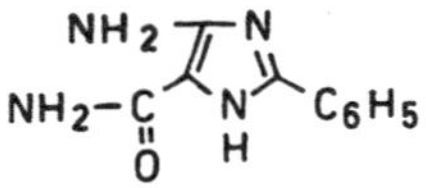	Walker carcinoma of rat	$\approx 10^{-3}$	reversible, competes with next substance	288
5(1-aziridinyl)-2,4-dinitrobenzamide 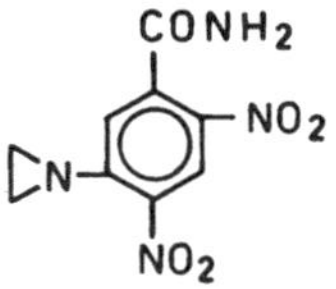	Walker carcinoma of rat and two human cell lines	$\approx 10^{-4}$	irreversible (alkylation?)	288
meso-α,β-diphenylsuccinate 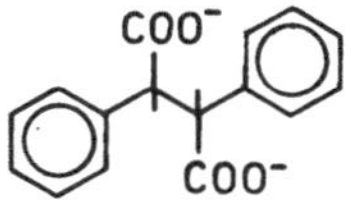	Ehrlich ascites tumor of mouse and mouse L cells	$\approx 10^{-3}$	(unspecific)	169

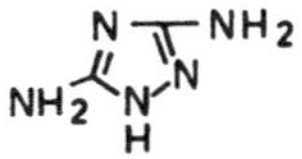

Table 10. (continued)

Type of compound	Inhibits enzyme of	I_{50} (M)[a]	Mode of action, where known	Ref.
c) miscellaneous aromatic compounds (continued)				
glutathione analogs with aromatic substituents	*E. coli*	3×10^{-3}	interaction with a glutathione binding site?	289
Cibacron blue F3GA	Ehrlich ascites tumor of mouse	2×10^{-4}	reversible, acts on nucleotide-binding subunit	67
pyridoxalphosphate	Ehrlich ascites tumor of mouse	$\approx 4 \times 10^{-4}$	reversible, competes with effector ATP	290
pyridoxalphosphate/$NaBH_4$	same	$\approx 10^{-3}$	irreversible, inactivates nucleotide-binding subunit	67, 284, 290, 305
d) chelators				
8-hydroxyquinoline and its 5-sulfonate	rat liver	$< 10^{-4}$		63
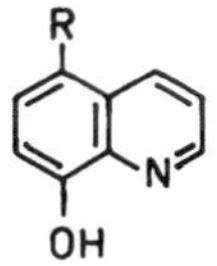 R=H, SO_3H	same		reversed by Fe(II)	64
	rabbit bone marrow	$< 10^{-4}$	prevented by Fe(II)	66
	E. coli		reactivation by Fe(II)	37
	inactive with *S. cerevisiae* and	–	–	50, 56
	calf thymus enzyme	–	–	
1,10-phenanthroline (o-phenanthroline)	rat liver	$< 10^{-4}$		63
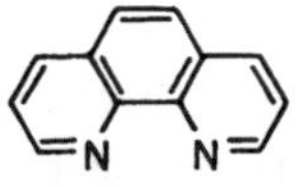	Novikoff hepatoma of rat		reversed by Fe(II)	61, 278
	embryonal chicken brain			84
	human tumor cells	4×10^{-4}		87
	Scenedesmus obliquus	10^{-4}	reversible without metal addition	48
	inactive with *S. cerevisiae* and probably *E. coli* enzyme	–	–	50, 37
bathophenanthroline disulfonate	rabbit bone marrow	$< 10^{-5}$	prevented by Fe(II)	66
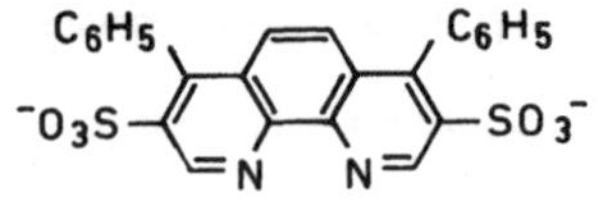	calf thymus		reversible without metal addition	56
	permeabilized mouse L cells	$\approx 10^{-4}$	reversible without metal addition	75
	(*inactive* with *E. coli* enzyme?)			37

(2-thenoyl)trifluoroacetone $C_4H_3S{-}C(=O){-}CH_2{-}C(=O){-}CF_3$	permeabilized mouse L cells	$\approx 10^{-4}$	reversible without metal addition	75
desferrioxamine (cf. Fig. 15)	H. Ep. 2 human carcinoma cells		prevented by Fe(II)	183
	human lymphocytes	$<10^{-3}$		184
	(*inactive* with *E. coli* enzyme?)			37
EDTA (ethylenediamine-N,N, N′,N′-tetraacetic acid)	Novikoff hepatoma of rat		reversed by Fe(III)	291
	rat liver	$\approx 10^{-2}$	(excess Mg(II) was present)	63
	H. Ep. 2 human carcinoma cells			183
	calf thymus	10^{-5}	reactivation by Fe(II) and Mn(II)	56, 346
	Ehrlich ascites tumor of mouse (only CDP reduction, ADP reduction was *stimulated*)	10^{-3}	partial reactivation by Mg(II)	134
	permeabilized mouse L cells (only CDP reduction, GDP reduction was *stimulated*)	3×10^{-3}	partial reactivation by Mg(II)	75
	Scenedesmus obliquus	10^{-2}		48
	E. coli		reversed by Mg(II)	33
	stimulates membrane-bound enzyme of *E. coli*	–		292
	Brevibacterium ammoniagenes	$<10^{-3}$		44
	inactive with *S. cerevisiae* and phage T 4 enzyme	–	–	52, 90
EGTA (ethyleneglycol-bis-(2-aminoethylether)-N,N,N′,N′-tetraacetic acid)	permeabilized mouse L cells (GDP reduction, CDP reduction was quite insensitive)	10^{-2}	Ca(II) complexation?	75

[a] The I_{50} figures are approximate, often extrapolated from tables, graphs, or K_i data. Where no figure is given, the inhibitor was not present in the assay but during dialysis

like a substrate, CDP, or the effector ATP strengthens this conclusion[182, 274, 278, 283]. An alternative interpretation could be that aromatic compounds act as effective inhibitors through their general capacity to participate in one-electron redox processes[180], in accord with the importance of radical scavenging in enzyme inactivation. The observed antagonism between inhibitors like 2,3-dihydro-pyrazoloimidazole or 4-methyl-5-amino-isoquinoline-1-carboxaldehyde thiosemicarbazone and hydroxyurea[294] is compatible with either possibility. In contrast, *E. coli* enzyme is rather insensitive towards the heterocyclic thiosemicarbazones[274, 278, 283], the aromatic hydroxamates[62], and other chelators[37]. It may lack the postulated hydrophobic binding site or the active center is less accessible, as has been discussed above (Table 8).

The α-(N)-heterocyclic carboxaldehyde thiosemicarbazones comprise the most active inhibitors of mammalian ribonucleotide reductases ($I_{50} \leq 10^{-7}$ M). Numerous compounds of this class have been studied[274, 276–279, 282, 293]. Their common conjugated, tridentate N-N-S ligand makes them powerful metal chelators so that initially they were believed to capture or mask enzyme-bound iron ions. However inhibition increases rather than is reversed by simultaneously added iron[278] and the ferrous chelates have now been recognized to be the actually inhibiting species[275, 279, 281, 283]. Accordingly, their potency is diminished by other iron-complexing agents[280, 294]. The molecular mechanism of action has very recently been elucidated[281]. It was shown that in an oxygen-dependent reaction the enzyme's free radical is destroyed, probably by a reactive species generated in the interaction of Fe(II) and O_2, analogous to the Fe(II)-bleomycin-oxygen-catalyzed degradation of DNA. This model explains the long known competition with dithiols[274, 278, 283] which can serve for restoration of the free radical under otherwise identical conditions[60]. The aromatic character of the inhibitor may facilitate selective binding to ribonucleotide reductase. Some other metal chelates like 1,10-phenanthroline copper[295, 296], pyridine-2-carboxaldehyde thiosemicarbazone copper[297], and even ferrous EDTA[298] generate oxygen radicals, and it is tempting to speculate whether this property might more generally apply to the inhibition of ribonucleotide reductases by chelators, which is not always accompanied by loss of metal ions from the enzymes[48, 56, 75].

Phenols (class b in Table 10) are well-known radical scavengers (one-electron donators) and this could explain enzyme inhibition produced by pyrogallol and catechol derivatives[62, 198]. Alternative to direct electron transfer, enzyme inactivation could proceed via oxygen radicals, generated by autoxidation of these compounds[299]. Aurintricarboxylic acid represents a particularly interesting case. This inhibitor of protein-nucleic acid interactions is the most efficient inhibitor of *E. coli* ribonucleotide reductase so far observed. Its active form is an oligomer of $\approx$6,000 dalton and bears a free radical itself[301, 302]. It is very reasonable that the combination of polyanionic nature, chelating ability, and radical content make aurintricarboxylic acid such a potent inhibitor.

Among the other structures, guanazole deserves mention as it behaves like an analog of hydroxyurea and has recently been recognized as radical scavenger, too[62]. The remaining compounds have been studied much less. The activity of iron-specific chelators and EDTA (class d), and reversal of their effects by iron salts point to the involvement of iron in eukaryotic ribonucleotide reductases. However, other metals have scarcely been tried for reactivation; with calf thymus enzyme manganese was nearly as effective as ferrous iron[56]. The relieving effect of iron[172, 283, 286] or synergistic action of extra chelators[289, 294] does not necessarily indicate iron removal as cause of inhibition but may be a matter of radical reactivation, as discussed on p. 68. Surprisingly, chelating agents specific for Fe(III) have found little attention.

Unspecific inhibition of ribonucleotide reduction is produced by compounds like pyridoxal phosphate, or the sulfonated anthraquinone-triazine dye, Cibacron blue. They interact, like in many enzymes, with nucleotide binding domains where pyridoxal phosphate becomes covalently linked to lysine, or in that the dye occupies the whole nucleotide fold. The latter interaction permits its use in affinity chromatography of ribonucleotide reductases[57, 67, 70, 79]. Likewise, EDTA is not a specific, nor a potent inhibitor; it may, for example, act by complexation of the structure-stabilizing Mg^{++} ions in native holoenzymes. However the iron-promoted radical regeneration process appears far more susceptible to interference from EDTA[346].

It is an unexpected result from the numerous inhibitor studies summarized here to see the iron (or other metal) ribonucleotide reductases so little impressed by the classical, metal chelating ability of these compounds. It is the unique free radical structure of the metalloenzymes that makes the proteins vulnerable to inactivation while the metals themselves appear much less attractive or accessible as target.

4. *Naturally Occurring Inhibitors*

Common natural inhibitors of ribonucleotide reduction are the negative allosteric effectors dATP, dTTP, and to a lesser extent, dGTP. Several less defined substances interfering with deoxyribonucleotide formation in vitro have been observed in extracts of yeast, wheat, and rodent cells[50, 53, 303–306]. Some of these inhibitory fractions had spectral or chemical properties resembling nucleotides[50, 53, 303] while others were tentatively identified as protein [304–306]. Cellular as well as synthetic RNA efficiently inhibits ribonucleotide reductase from mouse Ehrlich ascites tumor ($I_{50} = 5 \times 10^{-6}$ M). Reversible competition with either substrate or effector nucleotides was discussed whereas the physiologic significance of the findings remains open[307]. Some phenolic metabolites isolated from mushrooms, or L-Dopa derivatives possibly exert their cytotoxic action through inhibition of ribonucleotide reductase, yet this appears to be a case for pharmacology rather than for cellular regulation[198].

The activity of the proteinaceous inhibitors isolated from animal cells is inversely correlated with the proliferation rate of their source. Only recently was such an "inhibitor" characterized as complex mixture of enzymes like phosphatases and kinases that act on the substrate CDP and on the effector ATP[306]. The assumption is justified that this class of substances merely interferes with the assay and does not define specific cellular inhibitors of ribonucleotide reductase. Nevertheless it must be realized that ribonucleotide reductase activity is well coordinated with cell proliferation and is strictly controlled within the cell cycle. Hence a regulatory role of inhibitors besides the control of gene expression and fine tuning by allosteric effectors is a priori conceivable.

A more promising group of such compounds has been isolated from fungi like the freshwater mould *Achlya*[308]. Three highly modified nucleotides were partially characterized, of which compound HS 3 contains an ADP-sugar-X-glutamate moiety, covalently linked to another unit composed of UDP, mannitol, and four phosphates[309]. These nucleotides appear involved in the control and integration of a variety of cellular processes, among them membrane transport of nucleosides[310] and glucose[311], RNA synthesis[312], growth and sporulation[313]. In particular, they efficiently inhibit ribonucleotide reductase from *Achlya* and Chinese hamster cells ($I_{50} = 10^{-4}$ M) but are barely active with *E. coli* enzyme[107]. The appearance of HS 3 during the cell cycle coincides with the decline of ribonucleotide reductase activity and DNA synthesis[107]. A substance similar to HS 3 was found in Chinese hamster ovary cells and induced to high levels after growth inhibition by withdrawal of essential nutrients[314]. It inhibits ribonucleotide reduction in vitro without competition from the substrates and in vivo leads to cessation of RNA and DNA synthesis[314, 315]. A further function of HS 3 might be the integration of de novo and salvage pathways of nucleotide synthesis[315]. In conclusion, HS 3 can be seen as metabolic "sensor" comparable to the guanosine polyphosphates of prokaryotes for adaption of nucleic acid metabolism to nutritional stress, with ribonucleotide reductase as a main target.

We have found similar compounds in plants. Extracts from wheat seeds catalyze formation of deoxyribonucleotides only when prepared from aged grains, whereas fresh seeds are inactive in vitro[53]. It could be shown that this difference is due to the presence in fresh wheat of heat-stable, inhibitory material, which could serve the physiologic function to prevent untimely germination. One inhibitor fraction was partially characterized as guanosine-rich oligonucleotides of about 4,000 dalton. It inhibits CDP reduction by the ribonucleotide reductases of *E. coli* or *Scenedesmus obliquus* in an uncompetitive manner. Synthetic oligonucleotides of similar size mimic the action whereas polynucleotides or a dinucleoside phosphate, ApA, are inactive[316]. Some analogy is seen with the inhibition of mouse tumor enzyme by RNA[307]. As both the HS nucleotides and the wheat inhibitor are complex ribonucleotide derivatives, primary interaction with ribonucleotide reductases at one or more nucleotide sites with subsequent attachment to a more extended protein region, possibly through ionic forces, appears a reasonable model to explain uncompetitive inhibition. Besides final structure analysis it remains to be seen whether such modulation of enzyme activity by natural oligonucleotides occurs in more organisms.

E. Cellular Regulation

Ribonucleotide reduction is under strict control in each living cell, both on the time scale, during the cell cycle, and through spatial or functional compartmentation. Moreover, enzyme activity in vivo is modulated by allosteric effectors and metabolic conditions as described in previous sections. The immense efforts made by a cell to control synthesis, activity, and products or ribonucleotide reductases reflect their importance. Such control is not necessitated by the comparatively small energy demand of DNA synthesis; it is vital because enlarged or depleted intracellular deoxyribonucleotide pools lead to severely impaired, unbalanced growth and, even more serious, exert genetic effects[300, 348] which ultimately endanger the cell's identity.

1. Ribonucleotide Reduction and the Cell Cycle

The observation that ribonucleotide reductase activity always reflects the proliferative capacity of cells and organisms can be substantiated in synchronous cell cultures. Activity measured in extracts from cells in G_1 phase of the cell cycle is low or zero whereas a distinctive peak is seen in S phase[48, 51, 73, 86, 107, 317, 318]. The perfect correlation of DNA synthesis, ribonucleotide reductase activity, thymidylate synthase activity (not discussed here), and intracellular deoxyribonucleotide pools in light-dark synchronized algae, shown in Fig. 16, is a good example. The same "peak enzyme" pattern is observed in extracts from synchronous yeast, mouse, rat, and human cells as well as for the coenzyme B 12-dependent reductase of a unicellular cyanobacterium[319]. It can also be found when ribonucleotide reduction is assayed in permeabilized cells[83, 244, 320] or is estimated from the incorporation of labelled ribonucleosides into deoxyribonucleotide pools or DNA during the cell cycle[267]. In cells synchronized by metabolic blocks and then followed for enzyme activity for only one successive cycle it could be questioned whether the ribonucleotide reductase peak is indeed a cell cycle-specific event or merely reflects the resumption of growth after release from the block. However the rhythmic pattern in green algae growing synchronously for many generations in a natural light-dark regime confirms cell cycle specificity. It was recently shown in mouse lymphoma cells that the two ribonucleotide reductase subunits do not change in parallel during the cell cycle, with subunit M 2 exhibiting the larger variations[71]. However in Ehrlich ascites tumor cells slow-down of proliferation was correlated to a decrease of subunit M 1 activity, and in regenerating rat liver both subunits variied in a noncoordinate manner[69].

The initiation or arrest of enzyme synthesis responsible for the peak of ribonucleotide reductase is discussed below. Other factors which may modulate the activity in vivo or when measured in crude homogenates include allosteric control by the endogenous deoxyribonucleotides, the action of late S phase-specific inhibitors like the one found in *Achlya*, or the redox status of thioredoxin and glutaredoxin[9]; however the cell cycle dependence of these reactions is little known. It is therefore desirable to assay ribonucleotide reductase in preparations which have been subjected to at least one precipitation and dialysis step. While reliable cell cycle-dependent activity data are thus obtained, the absolute figures are frequently an order of magnitude too low to account for the cell's need of ribonucleotide reduction for DNA synthesis. This unsatisfactory condition is most likely a problem of "quinary enzyme structure" (see below) but is not felt to invalidate the accumulated evidence for tight correlation of ribonucleotide reduction and the cell cycle as a whole.

The initiation of ribonucleotide reductase activity when cells approach S-phase is suppressed by protein synthesis inhibitors (cycloheximide or chloramphenicol, respectively) and this is taken as evidence for de novo synthesis of the enzyme for each cell cycle[48, 51, 319]. Of course inhibition of protein synthesis will per se block cell cycle progress and hence could prevent posttranslational activation of the enzyme like covalent modification, generation of free radical, or assembly of a multienzyme aggregate for DNA replication. The mode of enzyme regulation cannot be recognized as long as studies of gene expression throughout the cell cycle at the level of transcription and/or translation are not feasible. The only information about control of ribonucleotide reductase synthesis has come from inhibition of DNA replication but it remains doubtful whether these experiments mirror the physiologic regulation or merely represent a kind of "SOS"-response. Depleting bacterial and eukaryotic cells for thymidine nucleotides, either by starvation of auxotrophs[321, 322], or with antimetabolites like fluorouracil and fluorodeoxyuridine[48, 51, 209, 321, 323] or methotrexate[48, 323] leads to significantly increased levels of ribonucleotide reductase. The enzyme overproduction is sensitive to inhibition

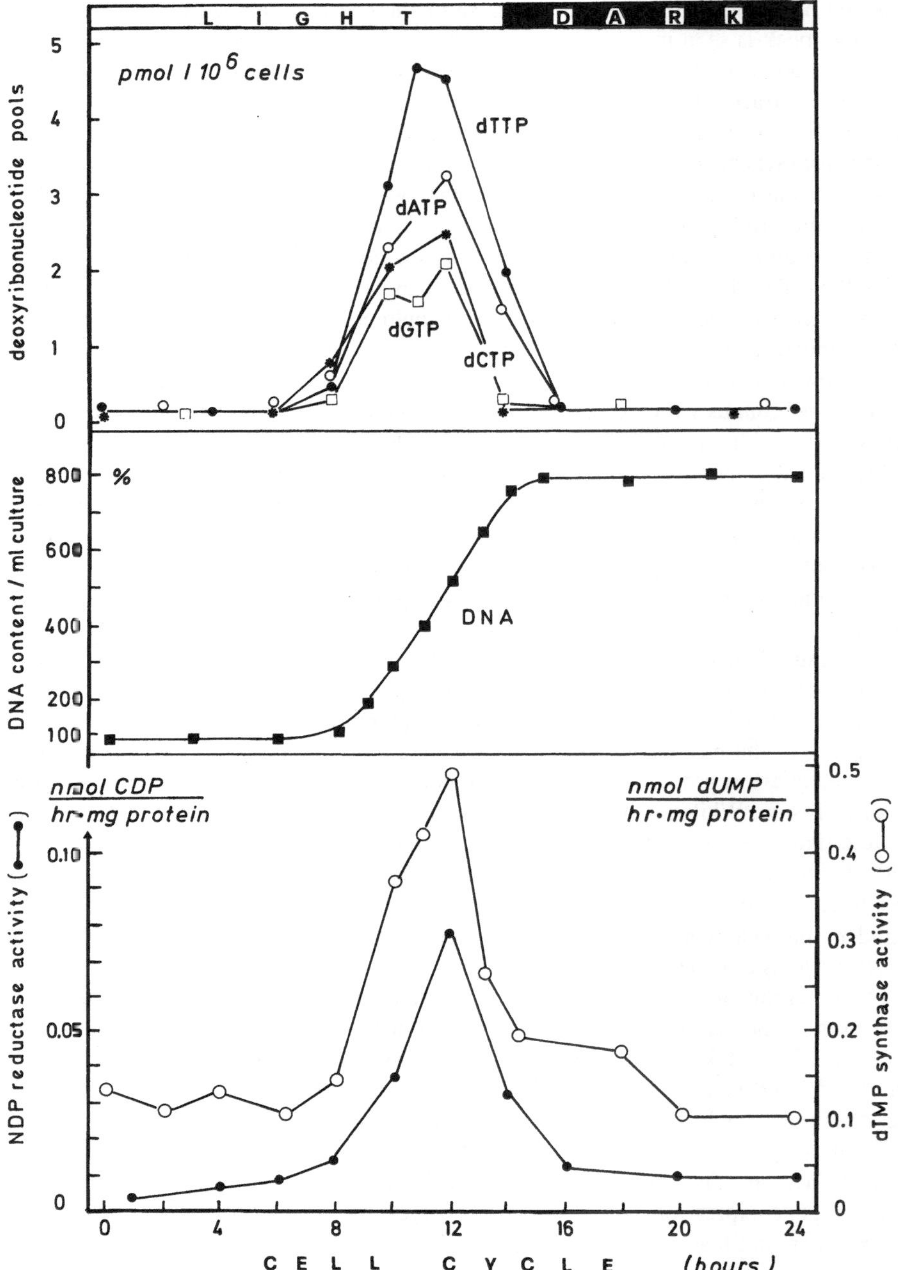

Fig. 16. Cell cycle dependence of intracellular deoxyribonucleotide pools (*top*) and the deoxyribonucleotide-synthesizing enzymes, ribonucleotide reductase and thymidylate synthase (*bottom*) in green algae (*Scenedesmus obliquus*) synchronized in a light-dark (14 : 10 h) regime[48, 209]

of protein synthesis, arguing against an activation of existing enzyme. In *E. coli* an increase of enzyme protein was also demonstrated by immunoelectrophoresis[30)]. Because of its magnitude (up to 10-fold, or more) and its occurence in synchronous cell cultures this increase of ribonucleotide reductase cannot merely be due to accumulation of S phase cells. (It is thus distinguished from the about twofold increase observed under DNA synthesis inhibition by hydroxyurea[198, 244)] or arabino-cytosine[198)], which arises from altered cell cycle kinetics). Depletion of cells for essential ribonucleotide reductase constituents like vitamin B 12, cobalt, or manganese, leads to analogous overproduction of (inactive) apoenzyme. This has been observed in *Rhizobium* species[27, 28)], in *Euglena gracilis* [26)], and in *Brevibacterium ammoniagenes*[44)], always accompanied by unbalanced growth and severely altered cell morphology. The obvious parallel of iron depletion may not be a practical possibility to induce iron-dependent ribonucleotide reductase in view of the multiple functions of this metal. It is remarkable that the conditions which derepress enzyme synthesis are so similar among all species examined, from bacteria[101)] to animals.

The decrease of enzyme activity after completion of DNA replication (Fig. 16), if measured in partially purified extracts to remove endogeneous inhibitors, must be attributed to arrest of enzyme synthesis, accompanied by inactivation or break-down of enzyme protein. Information on the termination of gene expression is indirect. As thymidylate depletion is a particularly efficient signal to stimulate ribonucleotide reductase synthesis above normal, it has been the basis for a model of regulation in which a thymidine nucleotide is to act as co-repressor[321, 323)]. This deduction, however, has not been verified experimentally. Thymidylate excess in Chinese hamster cells does neither abolish growth[324)] nor the increase of deoxyribonucleotide concentrations at the time of S phase[325)]. Deoxyadenosine also increases ribonucleotide reductase in HeLa cells in spite of only slightly decreased dTTP pool[85)]. In *E. coli* mutants with defective DNA replication, enzyme synthesis is stimulated even in presence of enlarged dTTP pools, as is true for inhibition of DNA synthesis by nalidixic acid[326)]. It appears unlikely that the regulatory signal for ribonucleotide reductase gene expression is one particular deoxyribonucleotide.

Speculations whether ribonucleotide reductase itself serves a regulatory or rate limiting role in DNA replication and cell cycle progress have repeatedly been based on the strict coupling of the processes, the in vitro activity below cellular needs, and on the minute amounts of intracellular free deoxyribonucleotides. No direct experimental evidence supports this view. The straightforward approach of supplying permeabilized G_1 phase cells with deoxyribonucleotides did not lead to earlier onset of DNA synthesis than in intact cells, arguing against control by DNA precursors for entry into S phase[217)]. An *E. coli* mutant with only 10% activity of ribonucleotide reductase subunit B 2 is not retarded in growth and independent of exogenous deoxyuridine under normal culture conditions[327)]. We conclude that in vitro activity is not an indicator of in vivo enzyme limitation.

If not the enzyme, one of its products might exert the regulatory function on DNA replication. The only deoxyribonucleotide without negative feedback action on ribonucleotide reductase, dCTP, has been advanced by Reichard to be involved. Addition of thymidine to hamster cells lowers the dCTP pool and concomitantly inhibits DNA synthesis even when dCTP has not become limiting as substrate[325, 328)]. Under these conditions no more initiation of replication takes place and a specific correlation of the two results has been claimed. Conversely, direct inhibition of DNA synthesis by aphidicolin[329] or 2′-deoxy-2′-azidocytidine[330] leads to selective decrease in cellular dCTP concentrations. As in other experiments thymidine-induced, severe depletion of dCTP pools was accompanied by only moderate inhibition of cell growth[324], final proof for regulatory role of dCTP in DNA synthesis has yet to come.

2. *An Enzyme Aggregate of DNA Synthesis?*

Catalysis of metabolic reaction chains by aggregates of enzymes offers several advantages to a cell: It enhances the overall reaction rate, channels the intermediates to yield high local concentration of an end product without cellular equilibration, and facilitates coordinate control of the whole process[331]. DNA replication is especially well suited for such organisation as it is restricted to distinct intracellular sites and utilizes specific precursors present in low intracellular concentrations and not involved in other metabolic reactions. Concentration gradients of deoxyribonucleotides, with high levels at the site of DNA replication and considerably lower overall concentrations have been demonstrated in studies of the incorporation of distal precursors into DNA of T 4 phage-infected *E. coli* cells[332] and an in vitro system from human lymphoblastoid cells[333]. Physical association of the precursor synthesizing enzymes with DNA polymerase is the most obvious explanation for such findings.

Multienzyme aggregates of that kind were indeed obtained in recent years in the laboratories of Elford, Mathews, Pardee, and others, and are summarized in Table 11. The aggregates are delicate entities, not to be compared with classical multienzyme complexes of well-defined composition like pyruvate dehydrogenase or fatty acid synthase; rather their constituent enzymes may be associated quite loosely, not amenable to common methods of isolation and size determination. Therefore the enzyme activities found in preparations from T 4 phage, bacteria, and mammalian cells (Table 11) may not be always identical with the physiologic, functional unit and are not yet suitable to define a general composition. However it is clearly seen that ribonucleotide reductase, thymidylate synthase, nucleotide kinases and DNA polymerase are associated in most instances, as expected for the function of channeling ribonucleotides into DNA. This enzyme apparatus has been named "replitase"[83]. Its supramolecular structure may represent an example of "quinary interactions"[343] among otherwise independent proteins.

The intracellular organisation of replitase is also only beginning to emerge. The bacterial complexes are apparently membrane-derived, and the rat liver enzyme aggregate was isolated from a postmicrosomal membran fraction, of unknown origin, too[339]. The aggregate of Chinese hamster cells is present only in S phase, but not in G_1 cells; in the former it is predominantly located in the nucleus whereas in G_1 separate enzyme activities are found mainly in the cytoplasm. Assembly of the enzymes within the nucleus was therefore proposed as the commiting step for onset of S phase, defining a new level of regulation in DNA synthesis and cell cycle progress[83, 273]. Nuclei from a population of thymus cells were also a good source for characterization of an enzyme complex[353]. In T 4 phage infected cells pyrimidine deoxyribonucleotides are synthesized with the start of DNA replication but in extracts the according enzyme activities can be measured earlier, the time lag indicating an assembly of the complex[340]. Earlier studies in which a certain amount of ribonucleotide reductase[63] or variable deoxyribonucleotide pools[252] were found in nuclei besides cytoplasmic localization can thus be explained as result of differ-

Table 11. Isolated multienzyme aggregates for DNA precursor synthesis and DNA replication

Source	Methods of characterisation	Constituents	Ref.
phage T4 infected *E. coli*	co-sedimentation on sucrose gradients gel chromatography ($M_r > 1.5 \times 10^6$) kinetic coupling analytical electrophoresis	ribonucleotide reductase, dCTPase/dUTPase, dCMP deaminase, thymidylate synthase, dCMP hydroxymethylase, dNMP kinase, NDP kinase, DNA polymerase, DNA topoisomerase	334, 335, 351
E. coli	differential centrifugation gel chromatography ($M_r = 2 \times 10^7$)	DNA, phospholipids, ribonucleotide reductase, thioredoxin, thymidylate synthase, DNA polymerase	336
E. coli	DNA-membrane-complex preparation	DNA, RNA, protein, lipids, ribonucleotide reductase	292
Streptococcus (Diplococcus) pneumoniae	DNA-membrane-complex preparation gel electrophoresis	DNA, RNA, protein, phospholipids, ribonucleotide reductase[a], deoxyribonucleotide kinases, DNA ligase, nucleases, DNA polymerase	337, 338
rat liver	co-sedimentation on sucrose gradients detergent and salt extraction electron microscopy	ribonucleotide reductase, thymidylate synthase, thymidine kinase, DNA polymerase	339
mouse carcinoma cells	co-sedimentation on sucrose gradients	thymidylate synthase, thymidine kinase, DNA polymerase	352
Chinese hamster fibroblasts	co-sedimentation on sucrose gradients kinetic coupling	nascent and template DNA, ribonucleotide reductase, thymidylate synthase, dihydrofolate reductase, thymidine kinase, NDP kinase, DNA polymerase	83, 273, 353
calf thymus	co-sedimentation on density gradients gel chromatography ($M_r = 5 \times 10^6$) gel electrophoresis electron microscopy	thymidine kinase, DNA methylase, DNA topoisomerase, DNA polymerase	353
human lymphoblastoid cells	gel chromatography ($M_r = 2 - 4 \times 10^5$) kinetic coupling	thymidylate synthase, thymidine kinase, dTMP kinase, dTDP kinase, DNA polymerase	333

[a] measured as ribonucleotide incorporation into DNA; low activity

ent stages of intracellular compartmentation. Finally the poor in vitro activity of most ribonucleotide reductase preparations is now understood as consequence of the disruption of enzyme aggregates during isolation, in accord with observations that the activity is severalfold higher in permeabilized[75, 244, 341] or gently lysed cells[336, 342] where it approaches the expected in vivo rates.

The enzyme studies described above are also compatible with a number of experiments in which incorporation of ribonucleotides into DNA has been shown to be more efficient than incorporation of deoxyribonucleotides[83, 335]. Functional compartmentation of DNA precursors is also observed in the utilization of deoxyuridine or thymidine for DNA which in *E. coli* or in eukaryotic cells occurs without prior equilibration with the "free" dTTP pool[344, 345]. On the other hand, in thymocytes purine deoxyribonucleosides are converted to nucleotides but are *not* utilized for DNA replication; however here they allosterically inhibit ribonucleotide reduction[255]. All these data agree with the existence of two different deoxyribonucleotide pools, one associated with the replitase complex and another independent pool of free deoxyribonucleotides.

The first pool probably cannot be quantified but is expected to contain equal amounts of all four DNA precursors. It is the second pool that represents the commonly measured intracellular dNTP concentrations which fluctuate during the cell cycle in parallel with ribonucleotide reductase[48, 86, 317], for example as shown in Fig. 16. These concentrations differ considerably among each other, with typical values found in the range from 5 to 100 pmol/10^6 cells in cultured mammalian cells in S phase[86, 249, 252, 267]. Such amounts would suffice for replication of at best 0.5% of the genome, demonstrating the indispensable role of ribonucleotide reduction for continuous DNA synthesis. These "free" deoxyribonucleotides are thought to sustain repair-type DNA synthesis or mitochondrial DNA replication as substrates. Moreover they will serve as the allosteric effector nucleotides that adjust ribonucleotide reductase activity and specificity to the needs for DNA replication, if the mechanisms of allosteric regulation unravelled in vitro (p. 54) also apply in vivo. Several investigations in which intracellular deoxyribonucleotide concentrations were manipulated by addition of exogenous deoxyribonucleosides or inhibitors[78, 79, 143, 144, 324, 325] are in fact in accord with the general model of Thelander and Reichard[8] for allosteric regulation of ribonucleotide reduction. One has to conclude that replitase-associated ribonucleotide reductase remains freely accessible for effectors from outside[33, 255].

The concepts of an enzyme complex for DNA precursor and DNA synthesis and of different deoxyribonucleotide pools should soon promote new insights into notoriously difficult to apprehend processes of cell biology. Ribonucleotide reduction which is pivotal in deoxyribonucleotide metabolism has been described above as an enzyme system of extreme complexity but now basically understandable function and origin. Analysis of its integration into supramolecular structure and regulation will, not so soon, open a new chapter.

We wish to thank many colleagues engaged in ribonucleotide reductase research for communicating unpublished observations to us. Our own work mentioned in the text is supported by Deutsche Forschungsgemeinschaft, Sonderforschungsbereich 103, "Zelldifferenzierung". Thanks are also due to all members of this research group for their continuing efforts, and patience.

F. References

1. Hammarsten, E., Reichard, P., Saluste, E.: J. Biol. Chem. *183,* 105 (1950)
2. Reichard, P., Baldesten, A., Rutberg, L.: ibid. *236,* 1150 (1961)
3. Larsson, A., Reichard, P.: ibid. *241,* 2540 (1966)
4. Reichard, P.: Eur. J. Biochem. *3,* 259 (1968)
5. Martin, D. W., Gelfand, E. W.: Annu. Rev. Biochem. *50,* 845 (1981)
6. Topal, M. D., Baker, M. S.: Proc. Natl. Acad. Sci. USA *79,* 2211 (1982)
7. Hogenkamp, H. P. C., Sando, G. N.: The enzymatic reduction of ribonucleotides, in: Structure and Bonding (eds.) Dunitz, J. D. et al., Vol. 20, p. 24, Berlin, Heidelberg, New York, Springer-Verlag 1974
8. Thelander, L., Reichard, P.: Annu. Rev. Biochem. *48,* 133 (1979)
9. Holmgren, A.: Curr. Topics Cell. Regul. *19,* 47 (1981)
10. Follmann, H.: Naturwiss. *69,* 75 (1982)
11. Blakley, R. L., Barker, H. A.: Biochem. Biophys. Res. Commun. *16,* 391 (1964)
12. Blakley, R. L.: Fed. Proceedings *25,* 1633 (1966)
13. Beck, W. S.: J. Biol. Chem. *242,* 3148 (1967)
14. Panagou, D., Orr, M. D., Dunstone, J. R., Blakley, R. L.: Biochem. *11,* 2378 (1972)
15. Hoffmann, P. J., Blakley, R. L.: ibid. *14,* 4804 (1975)
16. Abeles, R. H., Beck, W. S.: J. Biol. Chem. *242,* 3589 (1967)
17. Hogenkamp, H. P. C. et al.: ibid. *243,* 799 (1968)
18. Gleason, F. K., Hogenkamp, H. P. C.: Biochim. Biophys. Acta *277,* 466 (1972)
19. Gleason, F. K., Wood, J. M.: Science *192,* 1343 (1976)
20. Gleason, F. K.: FEMS Microbiol. Lett. *3,* 241 (1978)
21. Sando, G. N., Hogenkamp, H. P. C.: Biochem. *12,* 3316 (1973)
22. Gleason, F. K., Frick, T. D.: J. Biol. Chem. *255,* 7728 (1980)
23. Tsai, P. K., Hogenkamp, H. P. C.: ibid. *255,* 1273 (1980)
24. Gleason, F. K., Hogenkamp, H. P. C.: ibid. *245,* 4894 (1970)
25. Hamilton, F. D.: ibid. *249,* 4428 (1974)
26. Carell, E. F., Seeger, J. W., Jr.: Biochem. J. *188,* 573 (1980)
27. Cowles, J. R., Evans, H. J., Russell, S. A.: J. Bact. *97,* 1460 (1969)
28. Inukai, S., Sato, K., Shimizu, S.: Agric. Biol. Chem. *43,* 637 (1979); *44,* 1105 (1980)
29. Reichard, P., Moore, E. C.: Acta Chem. Scand. *17,* 889 (1963)
30. Eriksson, S., Sjöberg, B.-M., Hahne, S., Karlström, O.: J. Biol. Chem. *252,* 6132 (1977)
31. Platz, A., Sjöberg, B.-M.: J. Bact. *143,* 561 (1980)
32. Brown, N. C. et al.: Eur. J. Biochem. *9,* 561 (1969)
33. Thelander, L.: J. Biol. Chem. *248,* 4591 (1973)
34. Brown, N. C., Eliasson, R., Reichard, P., Thelander, L.: Eur. J. Biochem. *9,* 512 (1969)
35. Petersson, L. et al.: J. Biol. Chem. *255,* 6706 (1980)
36. Peters, J.: Nature *267,* 546 (1977)
37. Atkin, C. L., Thelander, L., Reichard, P., Lang, G.: J. Biol. Chem. *248,* 7464 (1973)
38. Stenkamp, R. E. et al.: Nature *291,* 263 (1981)
39. Sjöberg, B.-M., Loehr, T. M., Sanders-Loehr, J.: Biochem. *21,* 96 (1982)
40. Solbrig, R. M. et al.: J. Inorg. Biochem. *17,* 69 (1982)
41. Ehrenberg, A., Reichard, P.: J. Biol. Chem. *247,* 3485 (1972)
42. Sjöberg, B.-M., Reichard, P., Gräslund, A., Ehrenberg, A.: ibid. *252,* 536 (1977)
43. Sjöberg, B.-M., Reichard, P., Gräslund, A., Ehrenberg, A.: ibid. *253,* 6863 (1978)
44. Schimpff-Weiland, G., Follmann, H., Auling, G.: Biochem. Biophys. Res. Commun. *102,* 1276 (1981)
45. Willing, A., Follmann, H.: unpublished results (1982)
46. Auling, G.: unpublished results (1982)
47. Stutzenberger, F.: J. Gen. Microbiol. *81,* 501 (1974)
48. Feller, W., Schimpff-Weiland, G., Follmann, H.: Eur. J. Biochem. *110,* 85 (1980); Hofmann, R., Follmann, H.: unpublished results (1982)
49. Pries, M.: Ph. D. thesis, Marburg (1981)

50. Vitols, E., Bauer, V. A., Stanbrough, E. C.: Biochem. Biophys. Res. Commun. *41,* 71 (1970)
51. Lowdon, M., Vitols, E.: Arch. Biochem. Biophys. *158,* 177 (1973)
52. Lammers, M.: Ph. D. thesis, Marburg (1983)
53. Schimpff, G., Müller, H., Follmann, H.: Biochim. Biophys. Acta *520,* 70 (1978); Hoppe-S. Z. Physiol. Chem. *354,* 1299 (1973)
54. Hovemann, B., Follmann, H.: Biochim. Biophys. Acta *561,* 42 (1979)
55. Moore, E. C., Hurlbert, R. B.: ibid. *55,* 651 (1962)
56. Engström, Y., Eriksson, S., Thelander, L., Akerman, M.: Biochem. *18,* 2941 (1979)
57. Thelander, L., Eriksson, S., Akerman, M.: J. Biol. Chem. *255,* 7426 (1980)
58. Mattaliano, R. J., Sloan, A. M., Plumer, E. R., Klippenstein, G. L.: Biochem. Biophys. Res. Commun. *102,* 667 (1981)
59. Akerblom, L., Ehrenberg, A., Gräslund, A., Lankinen, H., Reichard, P., Thelander, L.: Proc. Natl. Acad. Sci. USA *78,* 2159 (1981)
60. Gräslund, A., Ehrenberg, A., Thelander, L.: J. Biol. Chem. *257,* 5711 (1982)
61. Moore, E. C.: Adv. Enzyme Regul. *15,* 101 (1977)
62. Kjøller Larsen, I., Sjöberg, B.-M., Thelander, L.: Eur. J. Biochem. *125,* 75 (1982)
63. Larsson, A.: ibid. *11,* 113 (1969)
64. Youdale, T., MacManus, J. P., Whitfield, J. F.: Can. J. Biochem. *60,* 463 (1982)
65. Hopper, S.: J. Biol. Chem. *247,* 3336 (1972)
66. Hopper, S.: Methods Enzymol. *51,* 237 (1978)
67. Cory, J. G., Fleischer, A. E., Munro, J. B.: J. Biol. Chem. *253,* 2898 (1978)
68. Cory, J. G., Fleischer, A. E.: Arch. Biochem. Biophys. *217,* 546 (1982)
69. Cory, J. G., Fleischer, A. E.: J. Biol. Chem. *257,* 1263 (1982)
70. Chang, C.-H., Cheng, Y.-C.: Cancer Res. *39,* 436 (1979)
71. Eriksson, S., Martin, D. W.: J. Biol. Chem. *256,* 9436 (1981)
72. Fuchs, J. A.: J. Bact. *130,* 957 (1977)
73. Turner, M. K., Abrams, R., Lieberman, I.: J. Biol. Chem. *243,* 3725 (1968)
74. Kuzik, B. A., Wright, J. A.: Enzyme *24,* 285 (1979)
75. Kucera, R., Paulus, H.: Arch. Biochem. Biophys. *214,* 114 (1982)
76. Kuzik, B. A., Wright, J. A.: Biochem. Genet. *18,* 311 (1980)
77. Fujioka, S., Silber, R.: J. Biol. Chem. *245,* 1688 (1970)
78. Meuth, M., Aufreiter, E., Reichard, P.: Eur. J. Biochem. *71,* 39 (1976)
79. Eriksson, S. et al.: J. Biol. Chem. *256,* 10184; 10193 (1981)
80. Elford, H. L., Freese, M., Passamani, E., Morris, H. P.: ibid. *245,* 5228 (1970)
81. Takeda, E., Weber, G.: Life Sci. *28,* 1007 (1981)
82. Peterson, D. M., Moore, E. C.: Biochim. Biophys. Acta *432,* 80 (1976)
83. Reddy, G. P. V., Pardee, A. B.: Proc. Natl. Acad. Sci. USA *77,* 3312 (1980); J. Biol. Chem. *257,* 12526 (1982)
84. Millard, S. A.: J. Biol. Chem. *247,* 2395 (1972)
85. Lin, A. L., Elford, H. L.: ibid. *255,* 8523 (1980)
86. Tyrsted, G., Gamulin, V.: Nucl. Acids Res. *6,* 305 (1979)
87. Chang, C.-H., Cheng, Y.-C.: Biochem. Pharmacol. *27,* 2297 (1978)
88. Dick, J. E., Wright, J. A.: J. Cell. Physiol. *105,* 63 (1980)
89. Yeh, Y.-C., Tessman, I.: J. Biol. Chem. *247,* 3252 (1972)
90. Berglund, O.: ibid. *247,* 7270 (1972); *250,* 7450 (1975)
91. Berglund, O., Sjöberg, B.-M.: ibid. *254,* 253 (1979)
92. Sahlin, M., Gräslund, A., Ehrenberg, A., Sjöberg, B.-M.: ibid. *257,* 366 (1982)
93. Eriksson, S., Berglund, O.: Eur. J. Biochem. *46,* 271 (1974)
94. Ponce de Leon, M., Eisenberg, R. J., Cohen, G. H.: J. gen. Virol. *36,* 163 (1977)
95. Henry, B. E., Glaser, R., Hewetson, J., O'Callaghan, D.: Virol. *89,* 262 (1978)
96. Langelier, Y., Buttin, G.: J. gen. Virol. *57,* 21 (1981)
97. Huszar, D., Bacchetti, S.: J. Virol. *37,* 580 (1981)
98. Lankinen, H., Gräslund, A., Thelander, L.: ibid. *41,* 893 (1982)
99. Mathews, C. K.: Exptl. Cell Res. *92,* 47 (1975)
100. Rima, B. K., Takahashi, I.: J. Gen. Microbiol. *107,* 139 (1978)
101. Filpula, D., Fuchs, J. A.: J. Bact. *139,* 694 (1979)
102. Yau, S., Wachsman, J. T.: Mol. Cell. Biochem. *1,* 101 (1973)

103. Iordan, E. P., Vorob'eva, L. I., Gaiman, V. I.: Microbiology (engl. Transl. of Mikrobiologiya) *44,* 544 (1975)
104. Kollárová, M., Perečkó, D., Zelinka, J.: Biológia (Bratislava) *35,* 907 (1980)
105. Gleason, F. K., Wood, J. M.: J. Bact. *128,* 673 (1976)
106. Sprengel, G., Follmann, H.: FEBS-Lett. *132,* 207 (1981)
107. Lewis, W. H., McNaughton, D. R., LéJohn, H. B., Wright, J. A.: Biochem. Biophys. Res. Commun. *71,* 128 (1976)
108. Feller, W., Follmann, H.: ibid. *70,* 752 (1976)
109. Noronha, J. M., Sheys, G., Buchanan, J. M.: Proc. Nat. Acad. Sci. USA *69,* 2006 (1972)
110. de Petrocellis, B., Rossi, M.: Develop. Biol. *48,* 250 (1976)
111. Swindlehurst, M., Berry, S. J., Firshein, W.: Biochim. Biophys. Acta *228,* 313 (1971)
112. Tondeur-Six, N., Tencer, R., Brachet, J.: ibid. *395,* 41 (1975)
113. Reichard, P.: J. Biol. Chem. *236,* 2511 (1961)
114. Miller, H. K., Balis, M. E.: Arch. Biochem. Biophys. *126,* 221 (1968)
115. Gordon, H. L., Fiel, R. J.: Cancer Res. *29,* 1350 (1969)
116. Diekert, G., Jaenchen, R., Thauer, R. K.: FEBS-Lett. *119,* 118 (1980); Pfaltz, A. et al.: Helv. Chim. Acta *65,* 828 (1982)
117. Dickman, S. R.: J. Mol. Evol. *10,* 251 (1977)
118. Fox, G. E. et al.: Science *209,* 457 (1980)
119. Gale, G. R. et al.: Cancer Res. *24,* 1012 (1964)
120. Beck, C. F., Neuhard, J., Thomassen, E.: J. Bact. *129,* 305 (1977)
121. Buetow, D. E., Mego, J. L.: Biochim. Biophys. Acta *134,* 395 (1967)
122. Goetz, G. H., Carell, E. F.: Biochem. J. *170,* 631 (1978)
123. Odmark, G.: Physiol. Plant. *25,* 158 (1971)
124. Brachet, J.: Nature *214,* 1132 (1967)
125. von Döbeln, U., Eckstein, F.: Eur. J. Biochem. *43,* 215 (1974)
126. Follmann, H., Hogenkamp, H. P. C.: Biochem. *10,* 186 (1971)
127. Ludwig, W., Follmann, H.: Eur. J. Biochem. *91,* 493 (1978)
128. Thelander, L. et al.: J. Biol. Chem. *251,* 1398 (1976)
129. Stubbe, J., Kozarich, J. W.: J. Amer. Chem. Soc. *102,* 2505 (1980); J. Biol. Chem. *255,* 5511 (1980)
130. Sjöberg, B.-M., Gräslund, A., Eckstein, F.: J. Biol. Chem. *258,* in press (1983)
131. von Döbeln, U., Reichard, P.: ibid. *251,* 3616 (1976)
132. Eriksson, S., Thelander, L., Akerman, M.: Biochem. *18,* 2948 (1979)
133. Chang, C.-H., Cheng, Y.-C.: Cancer Res. *39,* 5081; 5087 (1979)
134. Cory, J. G., Mansell, M. M.: ibid. *35,* 2327 (1975)
135. Ludwig, W., Follmann, H.: Eur. J. Biochem. *82,* 393 (1978)
136. Follmann, H. in: NMR Spectroscopy in Molecular Biology (ed. Pullman, B.) p. 323, Dordrecht, Boston, London, Reidel Publ. Co. 1978
137. Brinkley, S. A., Lewis, A., Critz, W. J., Witt, L. L., Townsend, L. B., Blakley, R. L.: Biochem. *17,* 2350 (1978)
138. Kent, R. J., Heidelberger, C.: Mol. Pharmacol. *8,* 465 (1972)
139. Holmgren, A.: Trends Biochem. Sci. *5,* 26 (1981)
140. David, S., de Sennyey, G.: Carbohydr. Res. *77,* 79 (1979)
141. Berglund, O.: J. Biol. Chem. *247,* 7276 (1972)
142. Vitols, E., Brownson, C., Gardiner, W., Blakley, R. L.: ibid. *242,* 3035 (1967)
143. Kummer, D. et al.: Z. Krebsforsch. *91,* 23 (1978)
144. Ross, D. D. et al.: Cancer Res. *41,* 4493 (1981)
145. Hunting, D., Henderson, J. F.: CRC Crit. Rev. Biochem. *13,* 325 (1982)
146. Chen, A. K. et al.: Biochem. *13,* 654 (1974)
147. Brown, N. C., Reichard, P.: J. Mol. Biol. *46,* 25 and 39 (1969)
148. v. Döbeln, U.: Biochem. *16,* 4368 (1977)
149. Eriksson, S., Caras, I. W., Martin, D. W., Jr.: Proc. Natl. Acad. Sci. USA *79,* 81 (1982)
150. Caras, I. W., Martin, D. W., Jr.: J. Biol. Chem. *257,* 9508 (1982)
151. Batterham, T. J., Ghambeer, R. K., Blakley, R. L., Brownson, C.: Biochem. *6,* 1203 (1967)
152. David, S., Eustache, J.: Carbohydr. Res. *20,* 319 (1971)
153. Follmann, H., Hogenkamp, H. P. C.: Biochem. *8,* 4372 (1969)

154. Stubbe, J., Ackles, D.: J. Biol. Chem. *255,* 8027 (1980)
155. Stubbe, J., Ackles, D., Sehgal, R. K., Blakley, R. L.: ibid. *256,* 4843 (1981)
156. Vitols, E., Hogenkamp, H. P. C., Brownson, C., Blakley, R. L., Connellan, J.: Biochem. J. *104,* 580 (1967)
157. Thelander, L.: J. Biol. Chem. *249,* 4858 (1974)
158. Kim, J. J., Abrams, R., Franzen, J. S.: Arch. Biochem. Biophys. *182,* 674 (1977)
159. Hamilton, J. A., Yamada, R., Blakley, R. L., Hogenkamp, H. P. C., Looney, F. D., Winfield, M. E.: Biochem. *10,* 347 (1971)
160. Tamao, Y., Blakley, R. L.: ibid. *12,* 24 (1973)
161. Orme-Johnson, W. H., Beinert, H., Blakley, R. L.: J. Biol. Chem. *249,* 2338 (1974)
162. Coffman, R. E., Ishikawa, Y., Blakley, R. L., Beinert, H., Orme-Johnson, W. H.: Biochim. Biophys. Acta *444,* 307 (1976)
163. Blakley, R. L., Orme-Johnson, W. H., Bozdech, J. M.: Biochem. *18,* 2335 (1979)
164. Bernhard, W. A., Hüttermann, J., Müller, A.: Radiation Res. *68,* 390 (1976)
165. Robins, M. J., Wilson, J. S.: J. Am. Chem. Soc. *103,* 932 (1981)
166. Babior, B. M.: Acc. Chem. Res. *8,* 376 (1975)
167. Tkáč, A., Omelka, L.: Org. Magn. Reson. *13,* 406 (1980); *14,* 109 (1980)
168. Sato, A., Cory, J. G.: Biosci. Rpts. *1,* 627 (1981)
169. Turner, M. K., Abrams, R., Lieberman, I.: J. Biol. Chem. *241,* 5777 (1966)
170. Krakoff, I. H., Brown, N. C., Reichard, P.: Cancer Res. *28,* 1559 (1968)
171. Elford, H. L.: Biochem. Biophys. Res. Commun. *33,* 129 (1968)
172. Moore, E. C.: Cancer Res. *29,* 291 (1969)
173. Timson, J.: Mutation Res. *32,* 115 (1975)
174. Anderegg, G., L'Eplattenier, F., Schwarzenbach, G.: Helv. Chim. Acta *46,* 1400 (1963)
175. Neilands, J. B.: Struct. and Bonding *1,* 59 (1966)
176. Mizukami, S., Nagata, K.: Coordin. Chem. Rev. *3,* 267 (1968)
177. Chatterjee, B.: ibid. *26,* 281 (1978)
178. Boyland, E., Nery, R.: J. Chem. Soc. (C) 354 (1966)
179. Ramsbottom, J. V., Waters, W. A.: J. Chem. Soc. (B) 132 (1966)
180. Bloobstein, S. H. et al.: Biochem. Pharmacol. *27,* 2939 (1978)
181. Elford, H. L., Wampler, G. L., van't Riet, B.: Cancer Res. *39,* 844 (1979)
182. Hards, R. G., Wright, J. A.: J. Cell. Physiol. *106,* 309 (1981)
183. Brockman, R. W., Shaddix, S., Laster, W. R., Jr., Schabel, F. M., Jr.: Cancer Res. *30,* 2358 (1970)
184. Hoffbrand, A. V. et al.: British J. Haematol. *33,* 517 (1976)
185. Neilands, J. B.: Annu. Rev. Biochem. *50,* 715 (1981)
186. Neuhard, J.: Biochim. Biophys. Acta *145,* 1 (1967)
187. Neuhard, J., Thomassen, E.: Eur. J. Biochem. *20,* 36 (1971)
188. Sinha, N. K., Snustad, D. P.: J. Bact. *112,* 1321 (1972)
189. Rosenkranz, H. S. et al.: Biochim. Biophys. Acta *114,* 501 (1966)
190. Rosenkranz, H. S., Carr, H. S.: Cancer Res. *30,* 1926 (1970)
191. Karlström, H. O.: Eur. J. Biochem. *17,* 68 (1970)
192. Rosenkranz, H. S., Carr, H. S.: J. Bact. *92,* 178 (1966)
193. Rosenkranz, H. S., Carr, H. S., Pollak, R. D.: Biochim. Biophys. Acta *149,* 228 (1967)
194. Rosenkranz, H. S., Jacobs, S. J., Carr, H. S.: ibid. *161,* 428 (1968)
195. Fuchs, J. A., Karlström, H. O.: Eur. J. Biochem. *32,* 457 (1973)
196. Fuchs, J. A., Karlström, H. O.: J. Bacteriol. *128,* 810 (1976)
197. Wachsman, J. T., Morgan, D. D.: ibid. *105,* 787 (1971)
198. Elford, H. L. et al.: Adv. Enz. Regul. *19,* 151 (1981)
199. Fishbein, W. N., Carbone, P. P.: J. Biol. Chem. *240,* 2407 (1965)
200. Dixon, N. E. et al.: Can. J. Biochem. *58,* 1323 (1980)
201. Holmes, M. A., Matthews, B. W.: Biochem. *20,* 6912 (1981)
202. Coletti-Previero, M.-A. et al.: Biochem. Biophys. Res. Commun. *107,* 465 (1982)
203. Rich, P. R. et al.: Biochim. Biophys. Acta *525,* 325 (1978)
204. Parrish, D. J., Leopold, A. C.: Plant Physiol. *62,* 470 (1978)
205. Henry, M.-F., Nyns, E.-J.: Sub-Cell. Biochem. *4,* 1 (1975)

206. Clarkson, A. B., Grady, R. W.: The use of hydroxamic acids to block electron transport. In: Chemistry and Biology of Hydroxamic acids, Kehl, H., ed. p. 130. Basel, S. Karger Verlag, 1982
207. Rode, W. et al.: J. Biol. Chem. *255,* 1305 (1980)
208. Boehm, T. L. J., Kreis, W., Drahovsky, D.: Biochim. Biophys. Acta *696,* 52 (1982)
209. Bachmann, B., Hofmann, R., Follmann, H.: FEBS-Lett. *152,* 247 (1983)
210. Young, C. W., Hodas, S.: Science *146,* 1172 (1964)
211. Adams, R. L. P., Lindsay, J. G.: J. Biol. Chem. *242,* 1314 (1967)
212. Wawra, E., Dolejs, I.: Nucl. Acids Res. *7,* 1675 (1979)
213. Erixon, K., Ahnström, G.: Mutation Res. *59,* 257 (1979)
214. Puschendorf, B., Grunicke, H.: Biochim. Biophys. Acta *272,* 16 (1972)
215. Woodcock, D. M., Adams, J. K., Cooper, I. A.: ibid. *696,* 15 (1982)
216. Cress, A. E., Gerner, E. W.: Biochem. Biophys. Res. Commun. *87,* 773 (1979)
217. Miller, M. R., Castellot, J. J., Jr., Pardee, A. B.: Biochem. *17,* 1073 (1978)
218. Castellot, J. J. et al.: J. Biol. Chem. *254,* 6904 (1979)
219. Hotta, Y., Stern, H.: J. Mol. Biol. *55,* 337 (1971)
220. Stern, H., Hotta, Y.: Mol. Cell. Biochem. *29,* 145 (1980)
221. Johnston, L. H. et al.: Exptl. Cell Res. *141,* 53 (1982)
222. Fujiwara, Y.: Biophys. J. *15,* 403 (1975)
223. Clarkson, J. M.: Mutation Res. *52,* 273 (1978)
224. Francis, A. A. et al.: Biochim. Biophys. Acta *563,* 385 (1979)
225. Downes, C. S., Collins, A. R. S.: Nucl. Acids Res. *10,* 5357 (1982)
226. Cameron, I. L., Jeter, J. R., Jr.: Cell Tissue Kinet. *6,* 289 (1973)
227. Ockey, C. H., Allen, T. D.: Exptl. Cell Res. *93,* 275 (1975)
228. Howell, S. H., Blaschko, W. J., Drew, C. M.: J. Cell Biol. *67,* 126 (1975)
229. Miyata, H., Miyata, M., Ito, M.: Cell Struct. Funct. *4,* 81 (1979)
230. Sheinin, R. et al.: Can. J. Biochem. *58,* 1359 (1980)
231. Ross, D. W.: Virchows Arch. B Cell Pathol. *37,* 225 (1981)
232. Singer, R. A., Johnston, G. C.: Proc. Natl. Acad. Sci. USA *78,* 3030 (1981)
233. Walters, R. A., Tobey, R. A., Hildebrand, E.: Biochem. Biophys. Res. Commun. *69,* 212 (1976)
234. Cress, A. E., Gerner, E. W.: Exptl. Cell Res. *110,* 347 (1977)
235. Ford, S. S., Shackney, S. E.: Cancer Res. *37,* 2628 (1977)
236. Navarrete, M. H., Pérez-Villamil, B., López-Sáez, J. F.: Exptl. Cell Res. *124,* 151 (1979)
237. Tomita, K., Plager, J. E.: Cancer Res. *39,* 4407 (1979)
238. Stancel, G. M., Prescott, D. M., Liskay, R. M.: Proc. Natl. Acad. Sci. USA *78,* 6295 (1981)
239. Gallwitz, D.: Nature *257,* 247 (1975)
240. Moll, R., Wintersberger, E.: Proc. Natl. Acad. Sci. USA *73,* 1863 (1976)
241. Nadeau, P., Oliver, D. R., Chalkley, R.: Biochem. *17,* 4885 (1978)
242. Shephard, E. A. et al.: FEBS-Lett. *140,* 189 (1982)
243. Navalgund, L. G. et al.: J. Biol. Chem. *255,* 7386 (1980)
244. Lewis, W. H., Kuzik, B. A., Wright, J. A.: J. Cell. Physiol. *94,* 287 (1978)
245. Kim, J. H., Gelbard, A. S., Perez, G.: Cancer Res. *27,* 1301 (1967)
246. Bello, L. J.: Exptl. Cell Res. *89,* 263 (1974)
247. Wright, J. A., Hards, R. G., Dick, J. E.: Adv. Enz. Regul. *19,* 105 (1981)
248. Ashman, C. R., Reddy, G. P. V., Davidson, R. L.: Somat. Cell Genet. *7,* 751 (1981)
249. Skoog, L., Nordenskjöld, B.: Eur. J. Biochem. *19,* 81 (1971)
250. Adams, R. L. P., Berryman, S., Thomson, A.: Biochim. Biophys. Acta *240,* 455 (1971)
251. Walters, R. A., Tobey, R. A., Ratliff, R. L.: ibid. *319,* 336 (1973)
252. Skoog, L., Bjursell, G.: J. Biol. Chem. *249,* 6434 (1974)
253. Plagemann, P. G. W., Erbe, J.: J. Cell. Physiol. *83,* 321 (1974)
254. Tattersall, M. H. N. et al.: Europ. J. clin. Invest. *5,* 191 (1975)
255. Scott, F. W., Forsdyke, D. R.: Biochem. J. *190,* 721 (1980)
256. Tyrsted, G.: Biochem. Pharmacol. *31,* 3107 (1982)
257. Streifel, J. A., Howell, S. B.: Proc. Natl. Acad. Sci. USA *78,* 5132 (1981)
258. Abe, I. et al.: Gann *72,* 337 (1981)

259. Coyle, M. B., Strauss, B.: Cancer Res. *30,* 2314 (1970)
260. Walters, R. A., Tobey, R. A., Hildebrand, C. E.: Biochim. Biophys. Acta *447,* 36 (1976)
261. Johnston, L. H.: Curr. Genet. *2,* 175 (1980)
262. Cress, A. E., Bowden, G. T.: Biochem. Biophys. Res. Commun. *102,* 845 (1981)
263. Radford, I. R., Martin, R. F., Finch, L. R.: Biochim. Biophys. Acta *696,* 145 (1982)
Radford, I. R. et al.: ibid. *696,* 154 (1982)
264. Scudiero, D., Strauss, B.: J. Mol. Biol. *83,* 17 (1974)
265. Walker, I. G., Yatscoff, R. W., Sridhar, R.: Biochem. Biophys. Res. Commun. *77,* 403 (1977)
266. Sugrue, S. P., Desesso, J. M.: Teratology *26,* 71 (1982)
267. Hordern, J., Henderson, J. F.: Can. J. Biochem. *60,* 422 (1982)
268. Borun, T. W. et al.: Cell *4,* 59 (1975)
269. Sheinin, R., Lewis, P. N.: Somat. Cell Genet. *6,* 225 (1980)
270. Hereford, L. M. et al.: Cell *24,* 367 (1981)
271. Adams, R. L. P., Abrams, R., Lieberman, I.: J. Biol. Chem. *241,* 903 (1966)
272. Chiu, R. W., Baril, E. F.: ibid. *250,* 7951 (1975)
273. Reddy, G. P. V.: Biochem. Biophys. Res. Commun. *109,* 908 (1982)
274. Moore, E. C., Booth, B. A., Sartorelli, A. C.: Cancer Res. *31,* 325 (1971)
275. Saryan, L. A. et al.: J. Med. Chem. *22,* 1218 (1979)
276. French, F. A. et al.: ibid. *17,* 172 (1974)
277. Mbertus, S., Filipovic, P.: Eur. J. Med. Chem. *17,* 145 (1982)
278. Moore, E. C. et al.: Biochem. *9,* 4492 (1970)
279. Sartorelli, A. C. et al.: Adv. Enz. Regul. *15,* 117 (1977)
280. Cory, J. G., Lasater, L., Sato, A.: Biochem. Pharmacol. *30,* 979 (1981)
281. Thelander, L., Gräslund, A.: J. Biol. Chem., *258,* 4063 (1983)
282. Agrawal, K. C. et al.: Cancer Res. *37,* 1692 (1977)
283. Preidecker, P. J. et al.: Mol. Pharmacol. *18,* 507 (1980)
284. Cory, J. G., Fleischer, A. E.: Cancer Res. *39,* 4600 (1979)
285. Wright, J. A., Hermonat, M. W., Hards, R. G.: Bull. Environm. Contam. Toxicol. *28,* 480 (1982)
286. Cory, J. G., Fleischer, A. E.: Cancer Res. *40,* 3891 (1980)
287. Saeki, T. et al.: J. Antibiot. *27,* 225 (1974)
288. Tisdale, M. J., Habberfield, A. D.: Biochem. Pharmacol. *29,* 2845 (1980)
289. Höög, J.-O. et al.: FEBS-Lett. *138,* 59 (1982)
290. Cory, J. G., Mansell, M. M.: Cancer Res. *35,* 390 (1975)
291. Moore, E. C., Reichard, P.: J. Biol. Chem. *239,* 3453 (1964)
292. Viswanathan, G., Noronha, J. M.: Biochim. Biophys. Acta *567,* 325 (1979)
293. Sartorelli, A. C., Agrawal, K. C., Moore, E. C.: Biochem. Pharmacol. *20,* 3119 (1971)
294. Cory, J. G., Sato, A., Lasater, L.: Adv. Enz. Regul. *19,* 139 (1981)
295. Que, B. G., Downey, K. M., So, A. G.: Biochem. *19,* 5987 (1980)
296. Marshall, L. E. et al.: ibid. *20,* 244 (1981)
297. Saryan, L. A. et al.: Biochem. Pharmacol. *30,* 1595 (1981)
298. Hertzberg, R. P., Dervan, P. B.: J. Am. Chem. Soc. *104,* 313 (1982)
299. Kappus, H., Sies, H.: Experientia *37,* 1233 (1981)
300. Kunz, B. A.: Environm. Mutagenesis *4,* 695 (1982)
301. González, R. G., Blackburn, B. J., Schleich, T.: Biochim. Biophys. Acta *562,* 534 (1979)
302. González, R. G., Haxo, R., Schleich, T.: Biochem. *19,* 4299 (1980)
303. Cory, J. G., Monley, M. F.: Biochem. Biophys. Res. Commun. *41,* 1480 (1970)
304. Elford, H. L.: Adv. Enz. Regul. *10,* 19 (1972)
305. Cory, J. G.: ibid. *17,* 115 (1979)
306. Ikenaka, K. et al.: Gann *72,* 8 (1981)
307. Cory, J. G.: Cancer Res. *33,* 993 (1973)
308. LéJohn, H. B. et al.: Biochem. Biophys. Res. Commun. *66,* 460 (1975)
309. McNaughton, D. R. et al.: Can. J. Biochem. *56,* 217 (1978)
310. Stevenson, R. M., LéJohn, H. B.: Can. J. Biochem. *56,* 207 (1978)
311. Goh, S. H., LéJohn, H. B.: ibid. *56,* 246 (1978)
312. McNaughton, D. R., Klassen, G. R., LéJohn, H. B.: Biochem. Biophys. Res. Commun. *66,* 468 (1975)

313. LéJohn, H. B. et al.: Can. J. Biochem. *56,* 227 (1978)
314. Lewis, W. H. et al.: J. Cell. Physiol. *93,* 345 (1977)
315. Goh, S. H., Wright, J. A., LéJohn, H. B.: ibid. *93,* 353 (1977)
316. Baumann, H.: Ph. D. thesis, Marburg (1983)
317. Nordenskjöld, B. A., Skoog, L., Brown, N. C., Reichard, P.: J. Biol. Chem. *245,* 5360 (1970)
318. Hwang, K. M., Murphree, S. A., Shansky, C. W., Sartorelli, A. C.: Biochim. Biophys. Acta *366,* 143 (1974)
319. Gleason, F. K.: Arch. Microbiol. *123,* 15 (1979)
320. Lewis, W. H., Wright, J. A.: Somat. Cell Genet. *5,* 83 (1979)
321. Biswas, C., Hardy, J., Beck, W. S.: J. Biol. Chem. *240,* 3631 (1965)
322. Beck, W. S., Hardy, J.: Proc. Nat. Acad. Sci. USA *54,* 286 (1965)
323. Elford, H. L. et al.: Cancer Res. *37,* 4389 (1977)
324. Hunting, D., Henderson, J. F.: Biochem. Pharmacol. *31,* 1109 (1982)
325. Bjursell, G., Reichard, P.: J. Biol. Chem. *248,* 3904 (1973)
326. Filpula, D., Fuchs, J. A.: J. Bact. *130,* 107 (1977); *135,* 429 (1978)
327. Fuchs, J. A., Neuhard, J.: Eur. J. Biochem. *32,* 451 (1973)
328. Reichard, P.: Federation Proc. *37,* 9 (1978)
329. Nicander, B., Reichard, P.: Biochem. Biophys. Res. Commun. *103,* 148 (1981)
330. Akerblom, L., Pontis, E., Reichard, P.: J. Biol. Chem. *257,* 6776 (1982)
331. Gaertner, F. H.: Trends Biochem. Sci. *3,* 63 (1978)
332. Mathews, C. K., Sinha, N. K.: Proc. Natl. Acad. Sci. USA *79,* 302 (1982)
333. Wickremasinghe, R. G., Yaxley, J. C., Hoffbrand, A. V.: Eur. J. Biochem. *126,* 589 (1982)
334. Mathews, C. K., North, T. W., Reddy, G. P. V.: Adv. Enz. Regul. *17,* 133 (1979)
335. Allen, J. R., Reddy, G. P. V., Lasser, G. W., Mathews, C. K.: J. Biol. Chem. *255,* 7583 (1980)
336. Lunn, C. A., Pigiet, V.: ibid. *254,* 5008 (1979)
337. Firshein, W.: J. Mol. Biol. *70,* 383 (1972)
338. Firshein, W.: J. Bact. *118,* 1101 (1973)
339. Baril, E., Baril, B., Elford, H., Luftig, R. B.: in: Mechanism and Regulation of DNA Replication (ed. Kolber, A. R., Kohiyama, M.), p. 275, Plenum Press, New York (1974)
340. Tomich, P. K., Chiu, C.-S., Wovcha, M., Greenberg, G. R.: J. Biol. Chem. *249,* 7613 (1974)
341. Warner, H. R.: J. Bact. *115,* 18 (1973)
342. Eriksson, S.: Eur. J. Biochem. *56,* 289 (1975)
343. McConkey, E. H.: Proc. Natl. Acad. Sci. USA *79,* 3236 (1982)
344. Pato, M. L.: J. Bact. *140,* 518 (1979)
345. Taheri, M. R., Wickremasinghe, R. G., Hoffbrand, A. V.: Biochem. J. *194,* 451 and *196,* 225 (1981)
346. Thelander, L., Gräslund, A., Thelander, M.: Biochem. Biophys. Res. Commun. *110,* 859 (1983)
347. Löffler, M., Schimpff-Weiland, G., Follmann, H.: FEBS-Lett. *156,* 72 (1983)
348. Huszar, D., Bacchetti, S.: Nature *302,* 76 (1983)
349. Gale, G. R.: J. Bact. *91,* 499 (1966)
350. Hand, R., Tamm, I.: in: Cell Cycle Controls (ed. Padilla, G. M., Cameron, I. L., Zimmermann, A.), p. 273, Academic Press, New York (1974)
351. Chiu, C.-S., Cook, K. S., Greenberg, G. R.: J. Biol. Chem. *257,* 15087 (1982)
352. Ayusawa, D. et al.: ibid. *258,* 48 (1983)
353. Noguchi, H., Reddy, G. P. V., Pardee, A. B.: Cell *32,* 443 (1983)
354. Stubbe, J. A.: Mol. Cell. Biochem. *50,* 25 (1983)

Metabolism of the Carcinogen Chromate by Cellular Constituents

Paul H. Connett and Karen E. Wetterhahn

Department of Chemistry, Dartmouth College, Hanover, New Hampshire 03755, USA

The redox chemistry of chromium(VI) is discussed with respect to the cellular metabolism of the carcinogen chromate in vivo. Possible sites for cellular reduction of chromium(VI) to chromium(III) are considered. The reactions of amino acids, ascorbic acid, carboxylic acids, thiol-containing molecules and other small molecules with chromate under physiological conditions are presented. In general only ascorbate and those molecules containing sulfhydryl groups are capable of easily reducing chromate at pH 7.4. Thus, in the cytoplasm, glutathione, cysteine and ascorbate are likely candidates to react with chromate. While most proteins are unreactive toward chromate, certain redox proteins are active in reducing chromate. The heme proteins hemoglobin and cytochrome P-450 possess chromate-reductase activity, whereas cytochrome c and myoglobin are inactive. The NADPH-dependent flavoenzymes glutathione reductase and NADPH-cytochrome P-450 reductase also possess chromate-reductase activity. However, the NAD(P)H enzymes, isocitrate dehydrogenase, glutamate dehyrogenase and malate dehydrogenase do not reduce chromate. Both microsomes and mitochondria possess chromate-reductase activity. The microsomal activity is accounted for by the NADPH-cytochrome P-450 reductase/cytochrome P-450 system. The enzyme(s) responsible for the mitochondrial reduction of chromate have not been identified. Chromium(VI) and its metabolite chromium(III) inhibit the normal activities of enzymes which bind chromium(III) or reduce chromate. The metabolism of chromate involves the generation of reactive intermediates which ultimately bind to cellular constituents and damage their function in the cell.

Structure and Bonding 54

Abbreviations

GSH	glutathione (reduced form)
GSSG	glutathione (oxidized form)
Cys	cysteine
PSH	penicillamine (reduced form)
PSSP	penicillamine (oxidized form)
BAL	2,3-dimercaptopropanol
Unithiol	2,3-dimercapto-1-propane sulfonic acid
DTT	dithiothreitol
NADH	nicotinamide adenine dinucleotide (reduced form)
NAD^+	nicotinamide adenine dinucleotide (oxidized form)
NADPH	nicotinamide adenine dinucleotide phosphate (reduced form)
$NADP^+$	nicotinamide adenine dinucleotide phosphate (oxidized form)
RBC	red blood cell
Hb	deoxyhemoglobin
Hb^+	methemoglobin
Mb	deoxymyoglobin
Mb^+	metmyoglobin
GSSG-R	glutathione reductase
Tris	tris(hydroxymethyl)aminomethane
BP	benzo[*a*]pyrene
ER	endoplasmic reticulum
mt	mitochondria

1 Introduction

The carcinogenicity of chromium(VI) compounds has been well-documented in epidemiological studies and animal tests[1]. Chromium(VI) compounds were mutagenic in bacterial and mammalian cell systems whereas chromium(III) compounds were not active mutagens in the same test systems[2]. In subcellular assay systems both chromium(III) and chromium(VI) decreased the fidelity of *E. Coli* DNA polymerase I[3]. The differences in activity of chromium(VI) compared to chromium(III) in cellular and subcellular systems have been explained in terms of the "uptake-reduction" model of chromate carcinogenicity[4,5].

The inactivity of chromium(III) in cellular short term assay systems is probably due to the fact that cell membranes appear to be quite impermeable to most chromium(III) complexes. However, it appears that heterocyclic aromatic ligands facilitate uptake of chromium(III) into cells since *cis*-dichlorobis(2,2′-bipyridine)chromium(III), *cis*-dichlorobis(1,10-phenanthroline)chromium(III) and *cis*-oxalatobis(2,2′-bipyridine)-chromium(III) were acitve mutagens in the *Salmonella typhimurium* histidine reversion assay[6]. Chromate readily crosses cell membranes and enters cells[4,5].

Once chromate traverses the initial barrier of the cell membrane it can react with a variety of cellular components at different sites within the cell (Fig. 1). Reduction of chromate to chromium(III) may take place in the cytoplasm, mitochondria, endoplasmic reticulum or nucleus. Redox active macromolecules and/or small molecules in the cytoplasm and organelles are potential candidates for possessing cellular chromate-reductase

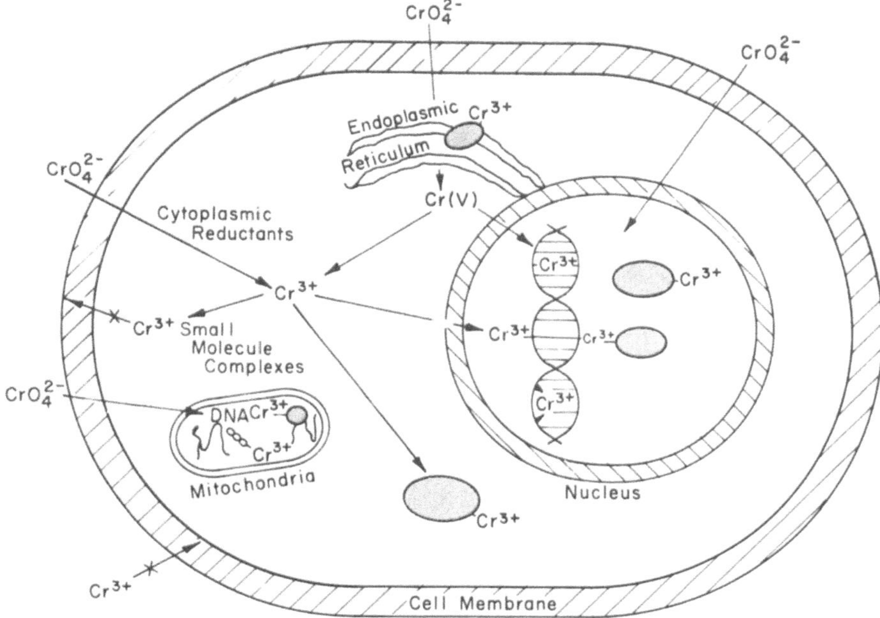

Fig. 1. The uptake-reduction model for chromate carcinogenicity. Possible sites for reduction of chromate include the cytoplasm, endoplasmic reticulum, mitochondria or the nucleus

activity. The reduction of chromate ultimately results in chromium(III) which binds to cellular small molecules, protein and DNA in a potentially damaging manner[5,7,8]. The chromium(III) complexes may inhibit normal cellular enzyme and gene activities. Chromate does not interact with DNA in the absence of metabolic reduction[8]. The nature of the cellular sites for chromate reduction and an assessment of their importance in the metabolism of chromate is the subject of this review.

2 Reduction of Chromium(VI) to Chromium(III)

2.1 *Thermodynamic Considerations*

Most oxidation reactions with chromium(VI) are carried out under acidic conditions[9]. The redox potential for chromium(VI) under acidic (pH = 0) conditions (Eq. 1) is + 1.33 V, whereas

$$Cr_2O_7^{2+} + 14\,H^+ + 6\,e^- = 2\,Cr^{3+} + 7\,H_2O \qquad E_0 = +1.33\ V \tag{1}$$

under alkaline (pH = 14) conditions (Eq. 2) the redox potential is − 0.12 V [10]. Thus, chromium(VI) is a much stronger oxidizing

$$CrO_4^{2-} + 4\,H_2O + 3\,e^- = Cr(OH)_3 + 5\,OH^- \qquad E_0 = -0.12\ V \tag{2}$$

agent under acidic conditions than under basic conditions.

Under physiological conditions the cellular pH is normally near 7.4. The stable forms of chromium in aerobic aqueous solutions near neutral pH (Fig. 2) are $HCrO_4^-$ and

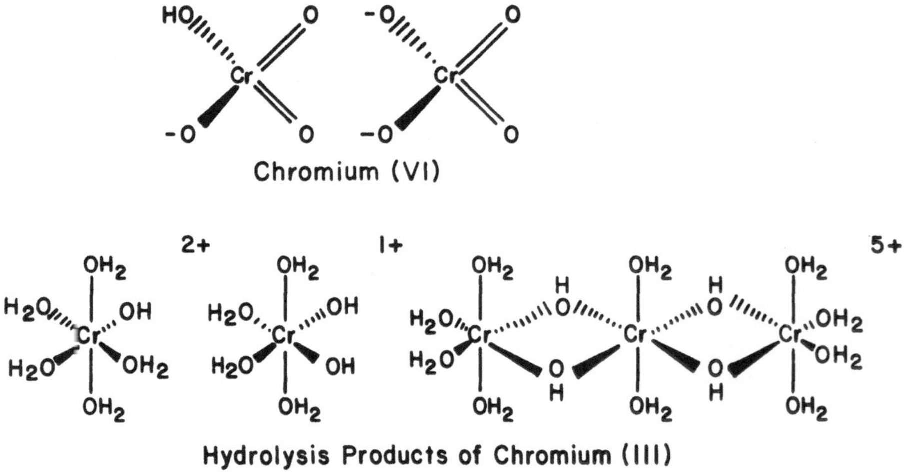

Chromium (VI)

Hydrolysis Products of Chromium (III)

Fig. 2. The forms of chromium(VI) and chromium(III) present in aqueous solution at neutral pH

CrO_4^{2-} for chromium(VI), and aquated $Cr_3(OH)_4^{5+}$, $Cr(OH)_2^{1+}$ and $Cr(OH)^{2+}$ for chromium(III)[11]. The relative amounts of the various species present in solution depend upon the actual pH and total chromium concentration. Extrapolation to pH 7.4 using the Nernst equation, gives redox potentials of + 0.34 Volts and + 0.52 Volts for the half reactions (1) and (2), respectively. Thus under neutral conditions Cr(VI) is a milder oxidizing agent than under acidic conditions. However, these potentials give only a rough idea of the driving force for Cr(VI) reduction within the cell, because it is unlikely that the final Cr(III) product will be either hexaquochromium(III) or $Cr(OH)_3$ but rather a Cr(III) complex with a cellular component. Thus cellular redox sites should be strong reducing moieties containing ligands which will stabilize the resulting Cr(III).

2.2 *Kinetic Factors*

The fate of Cr(VI) within the cell is not determined by thermodynamics alone. There are several cases, e.g., isocitrate (discussed below), of reducing agents which have redox potentials adequate to accomplish the reductions of Cr(VI) but which do so at an insignificant rate. Thus of key concern will be a consideration of the kinetic factors involved in the three electron conversion of Cr(VI) to Cr(III). To appreciate these factors we need to consider the structures of both Cr(VI) and Cr(III) and the structure, stability and lability of the intermediate valency states Cr(V) and Cr(IV).

2.2.1 Chromium(VI), d^0

In aqueous solution Cr(VI) can exist in three forms: $Cr_2O_7^{2-}$ (dominant species when pH < 7 and concentration $> 10^{-2}$ M), $HCrO_4^-$ (dominant species when pH < 6 and concentration $< 10^{-2}$ M), and CrO_4^{2-} (dominant species when pH > 6 at low concentration, and pH > 7 at high concentration)[11]. In each case the coordination about the Cr is tetrahedral. Under physiological conditions the dominant species is CrO_4^{2-}, chromate, but with a small and possibly significant percentage of $HCrO_4^-$ given by the following equilibrium[12]:

$$CrO_4^{2-} + H^+ \rightleftharpoons HCrO_4^- \,, \quad K = 7.94 \times 10^5 M^{-1} \tag{3}$$

Even a small percentage of $HCrO_4^-$ is significant since the oxygen atoms in this species are more labile to substitution than in CrO_4^{2-}[13].

2.2.2 Chromium(V), d^1

Cr(V) species have been observed via their narrow EPR signal at g ~ 1.98[14] and their maximum visible absorbance at wavelengths near 750 nm[15]. In addition, the existance of Cr(V) in reactions has been inferred from its ability to induce oxidation of I^- ion, which is accomplished only very slowly by Cr(VI)[16]. Salts of CrO_4^{3-} have been isolated from very basic solution and have been shown to contain a tetrahedral ion[17]. However, the EPR signals of some intermediates indicate an axial geometry and demonstrate that Cr(V) can have 5-fold coordination[18].

Usually aqueous solutions of Cr(V) are very unstable and undergo rapid disproportionation to Cr(VI) and Cr(III)[19]. However, Rocek and coworkers[20–24] have shown that oxalic acid and glycolic acid combine with Cr(V) to form relatively stable complexes. These same workers have more recently shown that 2-hydroxy-2-methylbutyric acid and citric acid form Cr(V) complexes sufficiently stable to permit storage of aqueous solutions of Cr(V) for extended periods of time (up to several weeks if stored frozen at $-20\,°C$)[19]. In further work they have been able to obtain crystals of potassium bis(2-hydroxy-2-methylbutyrate)oxochromate(V) monohydrate and to determine its crystal structure[25]. The geometry of the complex was intermediate between square pyramidal and trigonal bipyramidal. The ability of oxalic acid to form a stable complex with Cr(V) enables this compound to catalyze some Cr(VI) oxidations[26]. Square pyramidal complexes of oxalic acid and Cr(V) have been proposed on the basis of EPR studies[18].

2.2.3 Chromium(IV), d^2

Cr(IV) is a very unstable intermediate. According to Beattie and Haight[27], "Its reactions, including oxidation, reduction and disproportionation, are too fast to admit to its direct detection". However, K_2CrF_6 has been prepared and was shown to be isomorphous with K_2MnF_6[28]. Salts of tetrahedral CrO_4^{4-} are also known[29].

2.2.4 Chromium(III), d^3

All the Cr(III) complexes for which the structure has been determined show either a regular or slightly distorted octahedral geometry.

2.2.5 The Kinetic Labilities of Cr(V), Cr(IV) and Cr(III)

It is well known that Cr(III) complexes are kinetically inert to substitution[30]. Beck and Bardi[31] have shown that during chromate oxidations carried out in the presence of EDTA a Cr(III) EDTA complex is rapidly formed. This contrasts sharply with the direct reaction between Cr(III) and EDTA, which produces the complex only very slowly under otherwise identical conditions. These, and other ligand capture experiments[32], indicate that, unlike Cr(III), either or both Cr(V) and Cr(IV) are substitutionally labile. Cooper et al.[32] have performed ligand capture experiments which are consistent with the Cr(IV) species being more substitutionally labile than the Cr(V) species. They showed that with chromate oxidations initiated by VO^{2+} [a fast 1-electron reductant with Cr(IV) thought to be the immediate precursor of Cr(III)] the chromium captured a higher proportion of the non-redox active ligands SO_4^{2-}, N_3^- and $H_2PO_3^-$ than when comparable oxidations were initiated by $N_2H_5^+$ [a fast 2-electron reductant with Cr(V) thought to be the immediate precursor of Cr(III)].

While little is known about the actual rates or equilibria of Cr(V) and Cr(IV) substitution reactions, Hintze and Rocek[26] have been able to show from a study of the oxalic acid catalysis of chromic acid oxidation of tris(1,10-phenanthroline)iron(II) that the substitution reactions of Cr(V) and Cr(IV) must be at least an order of magnitude greater than $4 \times 10^5\ M^{-1}s^{-1}$.

2.2.6 *Rate-Limiting Steps*

Thus, in chromate reduction the overall change is from a labile tetrahedral Cr(VI) anion to an inert octahedral Cr(III) complex. The rate-limiting step for this process may be the step which involves the change in coordination number[9]. The least ambiguous results with respect to rate-limiting steps for chromate reactions carried out in acidic solution have been obtained for 1 electron reductants. Most reducing agents in this category, e.g., Fe^{2+} [33], VO^{2+} [34], NpO_2^+ [35], have the step Cr(V) to Cr(IV) as rate-limiting. However, there are other reductants, e.g., $Fe(CN)_6^{4-}$ [36], $Fe(phen)_3^{2+}$ [33], which have Cr(VI) to Cr(V) as their rate-limiting step. In these latter cases it would appear that it is a consequence of their kinetic inertness to substitution. The key point is that no study implicates Cr(IV) to Cr(III) as a rate-limiting step. Another feature which may be significant in the kinetics of conversion of Cr(VI) to Cr(III) is the availability of protons. As written in acid solution the reactions are:

$$CrO_4^{2-} + 8\,H^+ + 3\,e^- = Cr^{3+} + 4\,H_2O \qquad (4)$$

$$HCrO_4^- + 7\,H^+ + 3\,e^- = Cr^{3+} + 4\,H_2O \qquad (5)$$

Thus either 7 or 8 protons are required to convert Cr(VI) to Cr(III) in acidic solution reactions. In alkaline solution (Eq. 2) water is required for protonation of the chromate oxo groups.

We can predict that in the progression from Cr(VI) to Cr(III) as the electrons go into the d orbitals which are π-antibonding in character that the bonding between Cr and O will get progressively weaker and the basicity of the oxochromium species will increase. Thus we would expect proton availability to greatly affect the early stages of the reaction. Thus it is not surprising to find a proton concentration term appearing in the rate laws determined for interaction of chromate and several biological components, e.g., cysteine[37] and glutathione[38], carried out in acid solution. In physiological systems the protons required may come from cellular buffering components, the reductant itself or from water.

3 Cellular Sites for Redox Reactions with Chromate

From in vitro studies, possible reductants for Cr(VI) can range from small molecules and ions in the cytoplasm to complex membrane-bound enzyme systems in the endoplasmic reticulum. These reductants include certain amino acids and carboxylic acids, small peptides such as glutathione, components of electron transport systems in both the mitochondria and the microsomes, and proteins such as hemoglobin which while not normally functioning as electron transfer agents do contain redox sites.

4 Amino Acids

4.1 Cysteine

In 1958, Samitz and Pomerantz[39] showed that an ointment containing cysteine hydrochloride (2% in polyethylene glycol) was capable of blocking the effects of a 0.28% solution of dichromate applied to the skin of patients known to be subject to chromate dermatitis. In 1964 Samitz and Katz[40] showed that out of twenty amino acids tested, only cysteine and methionine were capable of reducing chromate after 48 h at pH 7 (phosphate buffer) and 37 °C. McCann and McAuley[37] studied the reaction between cysteine and chromate in perchlorate media in the pH range 1–1.7. Under these conditions they showed that the reaction proceded via the formation of a transient orange colored ($\lambda_{max} = 420$ nm, $\varepsilon_{max} = 1410 \pm 60\ M^{-1}cm^{-1}$) 1:1 chromate-thioester:

$$HOCrO_3^- + RSH \overset{K_1}{\rightleftharpoons} RS\text{–}CrO_3^- + H_2O \tag{6}$$

From their kinetic data they deduced the following rate law:

$$\frac{-d[Cr(VI)]}{dt} = \frac{K_1[cys][HCrO_4^-](k_1[H^+] + k_2[cys])}{1 + K_1[cys]} \tag{7}$$

This rate law is consistent with the following 2 pathway mechanism: the reduction of the chromate-thioester by acid

$$RS\text{–}CrO_3^- + H^+ \xrightarrow{k_1} Cr(V) \tag{8}$$

and the reduction of the thioester via reaction with a second molecule of cysteine:

$$RS\text{–}CrO_3^- + RSH \xrightarrow{k_2} Cr(IV) + RS\text{–}SR \tag{9}$$

They proposed that the purple species which accounted for about 60% of the Cr(III) product was a 1:1 Cr(III)-cystine complex (Fig. 3) with the cysteine bound through the N and O to the chromium ion.

The structure of blue crystals of sodium bis(L-cysteinate)chromium(III) dihydrate, **1**, which formed from a neutral aqueous solution of chromium(III) nitrate and L-cysteine, was found to be a slightly distorted octahedron with the Cr bound to two carboxylate oxygens and two amino nitrogens (mutually *cis*) and to two sulfur atoms (trans)[41]. This complex was found to be subject to acid hydrolysis which caused rapid cleavage of the chromium-sulfur bond and resulted in formation of the red-violet complex, $Cr(H_2O)_2(\text{L-cysteinate-O,N})_2^+$, **2**[42].

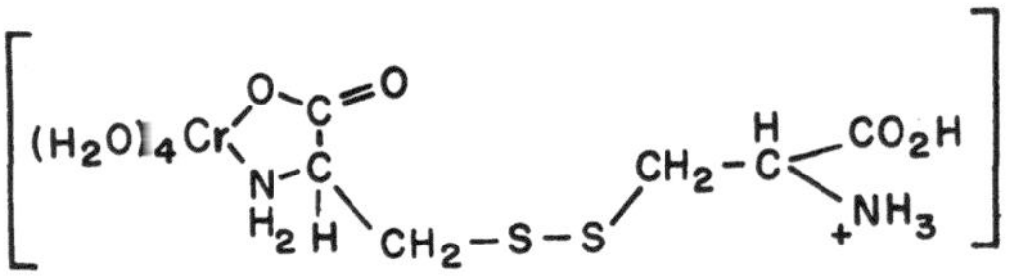

$$\left[(H_2O)_4Cr\begin{smallmatrix}O-C{=}O\\ N-C\\ H_2\ H\end{smallmatrix} CH_2-S-S-CH_2-\overset{H}{C}\begin{smallmatrix}CO_2H\\ NH_3^+\end{smallmatrix}\right]$$

Fig. 3. Structure proposed by McCann and McAuley[37] for the Cr(III) complex obtained in the reaction between Cr(VI) and cysteine in acidic media ($HClO_4$ = 0.02 – 0.1 M, I = 1.00 M, $NaClO_4$, 15–35 °C).

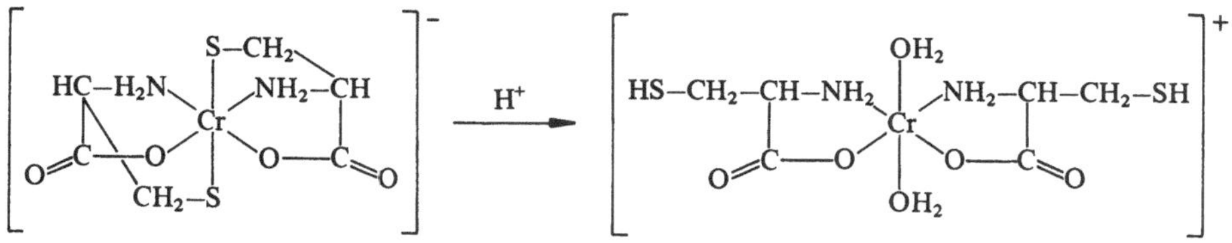

$$\mathbf{1}\ [\ldots]^{-} \xrightarrow{H^+} \mathbf{2}\ [\ldots]^{+} \qquad (10)$$

S–CH₂, HC–H₂N, NH₂–CH, Cr, O=C, O, O, C=O, CH₂–S — **1**

OH₂, HS–CH₂–CH–NH₂, NH₂–CH–CH₂–SH, Cr, O=C, O, O, C=O, OH₂ — **2**

In kinetic studies in our laboratory we have confirmed that cysteine efficiently reduces chromium(VI) at pH 7.4 (1 M Tris · HCl buffer) and 25 °C. Under identical concentration conditions cysteine reduced Cr(VI) more rapidly than either glutathione or penicillamine. Using a 66-fold molar (i.e. 22-fold redox equivalent) excess of cysteine, the reaction showed a first order dependence on chromate concentration. By varying the cysteine concentration over the range 16–130 × molar excess, the reaction showed a first order dependence in cysteine concentration. Thus at pH 7.4 the overall reaction appeared to show second order kinetics, with a second order rate constant of 75 ± 8 $M^{-1}min^{-1}$, at 25 °C.

If we compare these results with those of McCann and McAuley[37] at acid pH, we note that their rate law would approximate a second order rate law, of the form

$$\frac{-d[Cr(VI)]}{dt} = k_2[Cr(VI)][cys]$$

when $[H^+]$ is small and K_1 is large[9]. Their value of K_1 was 1030 ± 110 1 mol^{-1} at 25 °C, the value of k_1 was $9.4 \times 10^{-2}\ M^{-1}s^{-1}$, and the value of k_2 was $1.2 \times 10^{-2} M^{-1}s^{-1}$.

4.2 Methionine

Although Samitz and Katz[40] found that methionine was capable of reducing chromate after 48 h at 37° (pH = 7) our studies of the reaction between methionine and chromate at pH 7.4 (1 M Tris · HCl) showed no observable reaction at 25 °C. At 37 °C the reaction was extremely slow and certainly insignificant compared to cysteine. Chromium(III)-methionine complexes formed from reaction of chromium(III) with methionine contained coordinated $-NH_2$ and $-COO^-$ groups but no Cr–S bonds[43].

5 Ascorbic Acid

The reduction potential of ascorbic acid at pH 7 is + 0.08 V (see Table 1) and the redox change is shown in Eq. 11:

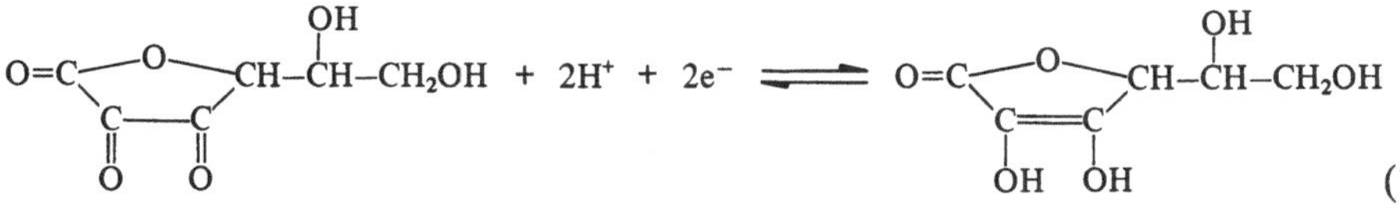

$$\text{dehydroascorbic acid (O=C–O–CH–CH(OH)–CH}_2\text{OH; C(=O)–C(=O))} + 2H^+ + 2e^- \rightleftharpoons \text{ascorbic acid (O=C–O–CH–CH(OH)–CH}_2\text{OH; C(OH)=C(OH))} \qquad (11)$$

Table 1. Standard reduction potentials, E_0' (pH 7, 25 °C), for some important cellular redox systems

Oxidant	Reductant	E_0' (Volts)
Acetate + CO_2	Pyruvate	− 0.70[a]
CO_2	Formate	− 0.42[a]
α-Ketoglutarate + CO_2	Isocitrate	− 0.38[b]
Uric Acid	Xanthine	− 0.36[a]
Acetoacetate	β-Hydroxybutyrate	− 0.35[b]
Pyruvate	Malate	− 0.33[a]
Cystine	Cysteine	− 0.32[a]
NAD^+	NADH	− 0.32[a]
$NADP^+$	NADPH	− 0.29[a]
Dehydrolipoate	Lipoate	− 0.23[a]
Glutathione (oxidized)	Glutathione (reduced)	− 0.23[a]
Dihydroxyacetone-Phosphate	Glycero-1-phosphate	− 0.22[c]
Pyruvate	Lactate	− 0.19[a]
Oxalacetate	Malate	− 0.17[a]
Hydroxypyruvate	Glycerate	− 0.15[c]
Pyruvate	Alanine/NH_4^+	− 0.13[a]
Glyoxylate	Glycolate	− 0.08[c]
Fumarate	Succinate	− 0.03[b]
Dehydroascorbate	Ascorbate	+ 0.06[a]
Ubiquinone	Ubiquinol	+ 0.10[a]

[a] from Ref. 118, [b] from Ref. 119, [c] from Ref. 120

In 1955, Rajka et al.[44)] demonstrated the ability of ascorbic acid to protect guinea pigs from the skin irritation caused by a 20% dichromate solution. In 1958, Samitz and Pomerantz[39)] showed that ascorbic acid (1% solution in polyethylene glycol) blocked the effect of a 0.25% dichromate solution on the skin of patients subject to chromate dermatitis. In an extension of this work, in 1962, Samitz et al.[45)] found that rats administered a lethal dose of potassium dichromate (130 mg/kg) via gastric intubation, followed within two hours by ascorbic acid, led to a 93% survival rate in the 28 rats tested. This and other work led to the use of ascorbic acid in the treatment of, and protection from, the toxic effects of chromate in industrial situations. This included the use of ascorbic acid impregnation of the material used in the respiratory masks of workers exposed to chromic acid mists[46)].

Evidence that ascorbic acid might be part of the body's natural defences against chromate poisoning was provided by Simavoryan[47)] who showed in 1971, that potassium dichromate given to rats, either orally or via subcutaneous injection, led to a decrease in the ascorbic acid levels in the kidney. Samitz et al.[48)] investigated the in vitro reaction of chromate and ascorbic acid and showed spectrophotometrically that ascorbate reduced Cr(VI) to Cr(III) with accompanying complex formation. Since they did not mention the presence of a buffer in their reaction mixtures ("hexavalent Cr" and "10% ascorbic acid") the reactions were probably carried out at acid pH.

In 1980, Baldea and Munteano[49)] investigated the kinetics of the reaction between chromate and ascorbic acid. They found that the reaction was first order with respect to both ascorbate and chromate and 0.7 order with respect to H^+ concentration. We have confirmed that at pH 7.4 (1 M Tris · HCl) and at 25 °C, that the reaction is first order in both chromate and ascorbate, and have found that the second order rate constant under these conditions was 36.1 ± 1.2 $M^{-1} min^{-1}$, which is intermediate between the values for cysteine (75.3 ± 6.2 $M^{-1} min^{-1}$) and penicillamine (20.9 ± 4.1 $M^{-1} min^{-1}$) obtained under the same conditions. Thus ascorbic acid must also be considered a candidate for reduction of chromate in the cell. Just how much chromate is reduced by ascorbic acid in vivo will probably hinge on its relative intracellular concentration compared with other reductants.

6 Carboxylic Acids

In 1963, Mali et al.[50)] showed that lactic acid reduced chromate at pH 4.3, and Samitz and Katz[40)] showed lactic acid reduction of chromate at pH 7 (in phosphate buffer) after incubation at 37 °C for 48 h. However, in our preliminary studies we found that in 1 M Tris · HCl at pH 7.4 and 25 °C, the reaction occurs very, very slowly and is certainly negligible compared to cysteine and ascorbic acid reactions carried out under identical conditions.

We have also looked at the reaction of chromate with a series of other carboxylates under these same conditions, including formate, oxalate, succinate, β-hydroxybutyrate, glycolate, malate, pyruvate and isocitrate. None of these candidate reductants reduced chromate at pH 7.4. These results emphasize the fact that chromate reduction under neutral or slightly alkaline conditions is under kinetic control, since all of the above carboxylates have favorable reduction potentials (see Table 1) and are thermodynamically capable of reducing chromate. Thus none of the carboxylic acids that have been examined, appears to be able to reduce chromate at pH 7.4 (1 M Tris · HCl) and 25 °C. These results again point to the significance of the SH group in chromate metabolism.

7 Thiol Containing Compounds

A key feature of the reaction between cysteine and chromate under acidic conditions (see Ref. 37 and Sect. 4.1) was the rapid formation of a chromate thioester. Similar thioesters have been observed in the reaction under acidic conditions between chromate and penicillamine, glutathione and cysteamine[51)], thiourea[52)], thiocyanate[53)] and thiosulfate[54)]. Although chromium is considered a "hard" acid and sulfur a "soft" base[55)], the equilibrium constants for the formation of these chromate thioesters are several orders of magnitude greater than the corresponding oxyesters[51)]. The readiness of the thiol group to interact with the chromate ion in addition to its strong reducing ability, makes it a particularly important entity in the metabolism of Cr(VI). In this section the reactions of chromate with a series of thiol containing compounds (both natural metabolites and model compounds) will be reviewed in order a) to see which thiol containing metabolites are likely candidates for in vivo reduction of chromate and b) to see what kind of environment around the thiol group most facilitates thiol reduction of chromate. The latter should help in predicting the amino acid sequences of enzymes most vulnerable to chromate attack.

7.1 Monothiols

7.1.1 Cysteine

This has already been discussed in Sect. 4.1. The second order rate constant of $75.3 \pm 6.2\ M^{-1}min^{-1}$ for the reaction of chromate with cysteine carried out at 25 °C and at pH 7.4 in 1 M Tris · HCl provides a useful reference against which we can compare the efficiency of other reductants (see Table 2). Observation of the visible spectrum (700 – 350 nm) during and after the reaction, under the same reaction conditions as above, and at various molar ratios ranging from 10 : 1 – 1 : 1 (cysteine : chromate), indicated that:

a) unlike the reaction carried out under acidic conditions[37], there was no obvious formation of thioester during the reaction, since there was no apparent peak or shoulder at 420 nm;

Table 2. *Second order rate constants determined for the reactions between chromate and some thiols, dithiols and ascorbate.*
In each case the rate constants were determined at pH 7.4 (1 M Tris · HCl) and 25 °C. The disappearance of chromate was monitored by observing the visible absorbance at 372 nm as a function of time using a CARY 219 UV-VIS spectrophotometer with the temperature controlled by a KR-4 Lauda circulating water bath. All the reagents used were obtained from Sigma Chemical Company, St. Louis, MO, except for the ascorbate and the chromate (Atomic Absorbance Reference) solution which were obtained from Fisher Scientific Company. The reagents were used without further purification and the concentration of the thiol solution determined using Ellman's reagent[121] The second order rate constants were determined by plotting the pseudo first order rate constants, obtained for a series of reactions in which the reductant was used in large excess, against reductant concentration. In each case $[Cr(VI)] = 3.7 \times 10^{-4}$ M and the reductant concentration was varied within the range 20–200 fold molar excess. The results are the average of two separate determinations.

Reductant	Second order rate constant ($M^{-1}min^{-1}$)
Cysteine	75.3 ± 6.2
Cysteamine	62.9 ± 3
Ascorbate	36.1 ± 1.2
Glutathione (first phase)	>26
Unithiol	26 ± 1.6
Penicillamine	20.9 ± 4.1
Dithiothreitol	17.3 ± 1.8
Mercaptoethanol	5.4 ± 0.2
Lipoic Acid	4.8 ± 0.5
Glutathione (final phase)	4.1 ± 1.5
2,3-Dimercaptosuccinic Acid	3.8 ± 0.2
Thiolactic Acid	3.1 ± 0.2

b) that the final product had peaks at 408 and 556 nm, with shoulders at about 460 and 600 nm respectively, indicating that the Cr was bound to N, O and S[57]. Moreover, the shift of the charge transfer band to lower energy, gave further evidence for Cr-S binding. These observations indicate that Cr(III) is captured by unoxidized cysteine during the reaction.

c) At a molar ratio of 3 : 1 (cysteine to chromate) a shoulder at 600 nm was clearly visible during the early stage of the reaction (within a minute of mixing) but this same reaction mixture showed no shoulder after it had been left for one day. This suggests that the unoxidized cysteine captured during the reaction is released slowly from the complex.

7.1.2 Penicillamine

Penicillamine (PSH), the dimethyl derivative of cysteine (structure given in Fig. 4), is not a natural metabolite but its reaction with chromate at pH 7 has been investigated because of its possible role as a detoxifying agent in chromate poisoning. In 1970 Sugiura, Hojo

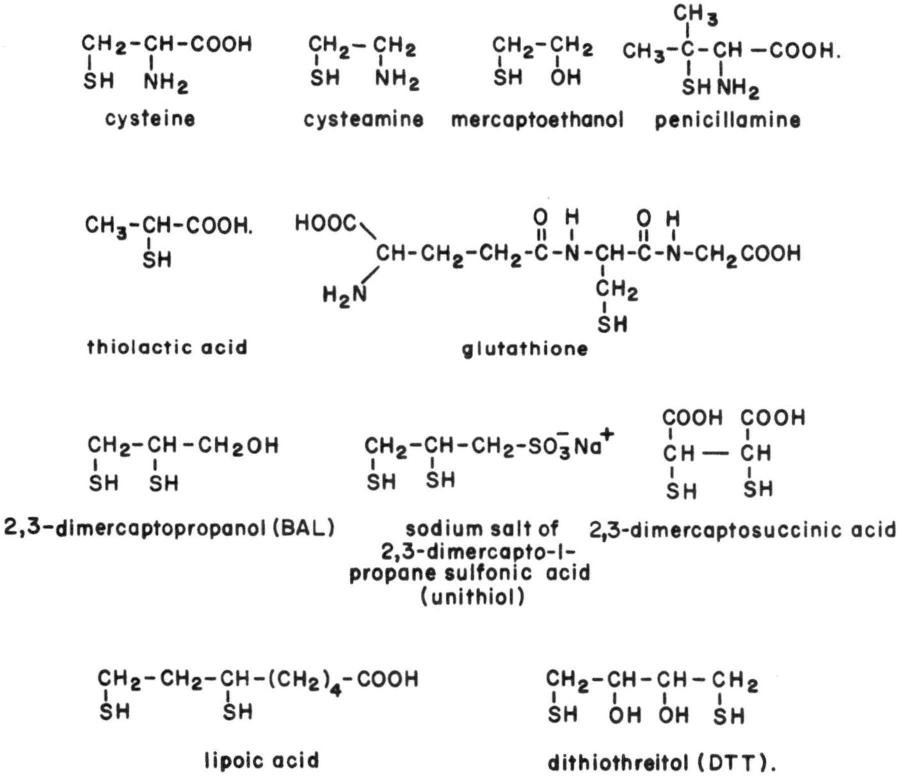

CH_2-CH-COOH (SH, NH_2)
cysteine

CH_2-CH_2 (SH, NH_2)
cysteamine

CH_2-CH_2 (SH, OH)
mercaptoethanol

CH_3-C(CH_3)-CH-COOH. (SH, NH_2)
penicillamine

CH_3-CH-COOH. (SH)
thiolactic acid

HOOC, H_2N > CH-CH_2-CH_2-C(=O)-N(H)-CH(CH_2-SH)-C(=O)-N(H)-CH_2COOH
glutathione

CH_2-CH-CH_2OH (SH, SH)
2,3-dimercaptopropanol (BAL)

CH_2-CH-CH_2-$SO_3^-Na^+$ (SH, SH)
sodium salt of 2,3-dimercapto-1-propane sulfonic acid (unithiol)

COOH COOH / CH — CH / SH SH
2,3-dimercaptosuccinic acid

CH_2-CH_2-CH-$(CH_2)_4$-COOH (SH, SH)
lipoic acid

CH_2-CH-CH-CH_2 (SH, OH, OH, SH)
dithiothreitol (DTT).

Fig. 4. Structures of the Thiols and Dithiols discussed in the text.

and Tanaka[56] reported a stoichiometry for the reaction carried out at pH 7 (0.03 M phosphate buffer) of PSH : CrO_4^{2-} = 6 : 1 and suggested a mechanism for the reaction in which 3 molecules of PSH reduce the Cr(VI) to Cr(III) and another 3 chelate with it to form a 1 : 3 complex. However, in 1977 this same group[57] in a more detailed study, reported a stoichiometry of 3 PSH : 1 CrO_4^{2-}, and suggested a mechanism in which 1 molecule of PSH reduces the Cr(VI) to Cr(V); 2 more molecules of PSH chelate to the Cr(V) and finally the Cr(V)-PSH complex is reduced to the corresponding Cr(III) complex by the solvent (0.2 M acetic acid/acetate buffer, pH 7). By measuring the initial rates of the reaction as a function of both penicillamine and chromate they determined the rate law to be

$$\text{rate} = k[\text{Cr(VI)}][\text{PSH}]$$

and their value for the second order rate constant at 30 °C and pH 7 was 0.549 ± 0.046 $M^{-1}s^{-1}$.

Our preliminary comparative study at pH 7.4 (1 M Tris · HCl) and 25°, also indicated a reaction first order in chromate and first order in PSH and gave us a value for the second order rate constant of 0.348 ± 0.063 $M^{-1}s^{-1}$. However, from the kinetic study by McAuley and Olatunji[38] for the reaction carried out under acid conditions, the above rate law would appear to be the simplification of the more complex rate law shown in Eq. 12:

$$\text{Rate} = \frac{K[\text{RSH}][\text{Cr(VI)}](k_1[H^+] + k_2[\text{RSH}])}{1 + K[\text{RSH}]} \tag{12}$$

This rate law and the mechanism proposed to explain it are analogous to that given for the chromate-cysteine reaction carried out in acid media: namely, the rapid formation of a thioester complex followed by its reduction via two alternative pathways, one involving a proton and the other a second molecule of penicillamine This is illustrated as follows:

$$\text{PSH} + \text{HOCrO}_3^- \overset{K}{\rightleftharpoons} \text{PSCrO}_3^- + H_2O \tag{13}$$

$$\text{PSCrO}_3^- + H^+ \xrightarrow{k_1} \text{Cr(V)} \tag{14}$$

$$\text{PSCrO}_3^- + \text{PSH} \xrightarrow{k_2} \text{Cr(IV)} + \text{PSSP} \tag{15}$$

The values of K, k_1 and k_2 determined at 25 °C over the pH range 1–1.7 in perchlorate media were as follows: K = 700 ± 40 M^{-1}, k_1 = 5.7 ± 1 × 10^{-2} $M^{-1}s^{-1}$ and k_2 = 14.3 ± 1 × 10^{-2} $M^{-1}s^{-1}$.

The crystal structure of L-histidinato-D-pencillaminatochromium(III) monohydrate has been determined[58]. The complex was synthesized from chromium(III) chloride, L-histidine and D-penicillamine in aqueous solution. The chromium(III) was coordinated to the N, S, and O of penicillamine and the N, O and imidazole (N) of histidine in an approximately octahedral geometry. Thus, both cysteine and penicillamine form complexes with Cr(III) which contain Cr-S bonds.

7.1.3 Glutathione

Glutathione is a tripeptide formed from glutamic acid (peptide bond from γ carboxyl group to cysteine), cysteine and glycine: its structure is given in Fig. 4. Glutathione is the most abundant naturally occurring low molecular weight thiol. It occurs in all mammalian cells and its intracellular concentration ranges from 0.8 to 8 mM[59]. Glutathione occurs in two oxidation states: reduced glutathione (GSH) and oxidized glutathione (GSSG) and the redox reaction is:

$$GSSG + 2H^+ + 2e \rightarrow 2GSH \quad E_0' = -0.24 \text{ V at pH } 7 \tag{16}$$

In 1958 Samitz and Pomerantz[39] described the use of glutathione (2% solution in polyethylene glycol) to protect skin from chromate attack. In 1960 Prins[60] in an investigation of the reaction between chromate and hemoglobin, observed the reaction between chromate and glutathione. He described the formation of a green complex which was retained on an anion exchange column. In 1972, Sugiura et al.[56] gave the spectral details of a chromium(III) GSH complex, which they obtained by heating Cr(III) with glutathione. The absorbance maxima of a resulting Cr(III)-glutathione complex were found to be 575 nm, 432 nm, and 270 nm. However, our studies with gluthathione resulted in purple glutathione complexes at neutral pH. These purple complexes are prepared by the reaction between Cr(VI) and GSH, or between Cr(III) and either GSH or GSSG, and show absorbance maxima different from those obtained by Sugiura et al.[56]. For all three products the visible absorbance maxima fall in the ranges 550 ± 5 nm and 405 ± 10 nm. The green complex observed by Prins may be a chromium phosphate complex, as we have found that if the Cr(VI) GSH reaction is carried out in phosphate buffer, that the purple complex is not formed and the product solution has very similar spectral characteristics to a solution prepared by adding Cr(III) salts to phosphate buffer. We also suspect that Sugiura et al.[56] must have used a large excess of glutathione in the preparation of their complex. The variation in our values of absorbance maxima for the species prepared by GSH reduction of Cr(VI) may be due to varying degrees of polymerization of the product depending upon the exact reaction conditions.

McAuley and Olatunji[38, 51] have investigated the kinetics of the reaction between glutathione and chromate in acid perchlorate media and they deduced the following rate law:

$$-\frac{d[Cr(VI)]}{dt} = \frac{K[Cr(VI)][GSH](k_1[H^+]^2 + k_2[GSH])}{1 + K[GSH]} \tag{17}$$

This rate law is analogous to that obtained for the reaction between chromate and both cysteine[37] and penicillamine[38] in acid perchlorate media, and again suggests two alternative breakdown pathways for a rapidly formed thioester. However, in the case of glutathione the kinetic data indicate that two protons are involved in the acid catalyzed pathway:

$$HOCrO_3^- + GSH \overset{K}{\rightleftharpoons} GSCrO_3^- + H_2O \tag{18}$$

$$GSCrO_3^- + 2H^+ \xrightarrow{k_1} Cr(V) \tag{19}$$

$$GSCrO_3^- + GSH \xrightarrow{k_2} Cr(IV) + GSSG \quad (20)$$

McAuley and Olatunji found the following values for the constants at 25 °C: K = 1436 ± 50 M^{-1} [51)], k_1 = 25 ± 4 × $10^{-2} M^{-2} s^{-1}$ [38)]. Moreover, from the variation of k_2 with temperature they determined the entropy of activation to be − 40 ± 5 cal $K^{-1}mol^{-1}$ for pathway 2. Such a large negative entropy value is indicative of a highly associative transition state and they suggested the possibility of the transition state illustrated in Fig. 5. Such a state would facilitate disulfide formation.

Our preliminary kinetic studies at pH 7.4 (1 M Tris · HCl or 1 M Tris · perchlorate) at 25 °C, indicate biphasic or even multiphasic kinetics; with a fast initial phase followed by a much slower second phase. The second phase is first order in chromate and appears to be first order in GSH and the second order rate constant for this second phase is 4.1 ± 1.5 $M^{-1}min^{-1}$. A minimum estimate of the rate constant for the first phase is 26 $M^{-1}min^{-1}$.

This is the only chromate-thiol reaction which gives this biphasic (or multiphasic) kinetics under our standard reaction conditions. The formation of Cr(III) (absorbance at 550 nm) is commensurate with the disappearance of Cr(VI) (absorbance at 372 nm) and the thioester appears only as a slight shoulder to the 372 nm peak. Preliminary investigation of the product formed in the reaction between Cr(VI) and GSH at pH 7.4 at a molar ratio of 1 : 3 indicates that a) it contains no free SH group, b) it contains no Cr-S bond, c) it contains Cr bound to both N and O, d) GSSG and not GSH is the ligand, e) the ratio of Cr : GSSG is 1 : 1, f) the charge on the complex is greater than 2 −, and g) the complex is polymeric although how rapidly the polymerization takes place has not been determined.

7.1.4 *Other Monothiols*

In comparative kinetic studies we have investigated the reaction between chromate and cysteamine, mercaptoethanol and thiolactic acid, at pH 7.4 (1 M Tris · HCl) and 25 °C. All the reactions appear to be first order with respect to both chromate and thiol. The second order rate constants are listed in Table 2. The rates of chromate reduction for the various monothiols follows the order: cysteine > cysteamine > glutathione (fast phase) > penicillamine > mercaptoethanol > glutathione (slow phase) > thiolactic acid.

7.2 *Dithiols*

The transition state proposed by McAuley and Olatunji[38)] (see Sect. 7.1.3) in which two thiol groups coordinate at adjacent sites on the Cr(VI) ion prior to disulfide formation

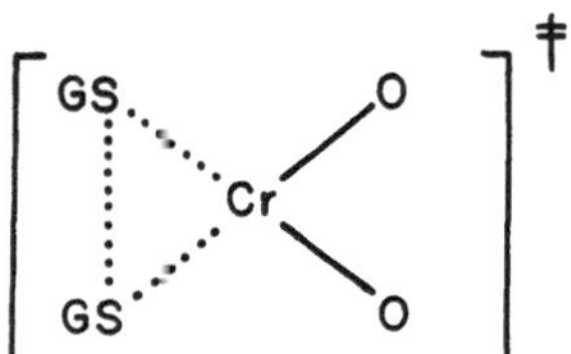

Fig. 5. Structure of the transition state proposed by McAuley and Olatunji[38)] for the reaction between chromate and thiols in acidic media ($HClO_4$ = 0.02 − 1.00 M, I = 1.00 M, $NaClO_4$).

makes the study of compounds containing two thiol groups particularly interesting. If ring closure is thermodynamically favorable, the possibility exists for much faster reactions with chromate than with single thiols. It would be useful therefore to have information on the reaction between chromate and dithiols in which the two thiol groups are separated by 2, 3, 4 and possibly more C atoms.

7.2.1 2,3-Dimercaptopropanol (BAL)

Historically the first chromate dithiol reaction studied was that with BAL (British Anti-Lewisite) a chelating agent used in the treatment of arsenic(III) (Lewisite) and other cases of heavy metal poisoning[61].

In 1946, Braun et al.[62] investigated the effect of injecting a 3% solution of BAL into the gluteal muscles of rabbits after they had received a toxic dose of a series of metal compounds. They found that the fatal dose of chromate was increased by BAL to about 1½ times that for the untreated controls. In 1953, Cole[63] demonstrated improvement in patients suffering from chrome dermatitis when they were treated with a 3% solution of BAL in zinc paste.

7.2.2 Unithiol (2,3-Dimercapto-1-propane Sulfonic Acid)

Unithiol (see Fig. 4 for structure) is the sodium salt of the sulfonic acid derivative of BAL. The compound is thus more soluble in water and is a crystalline solid at room temperature. It has its two thiol groups separated by 2 carbon atoms and thus if it forms a bisthioester with chromate, it would contain a 5-membered ring, and if the oxidation of such a thioester proceded intramolecularly, the disulfide bond formed would be contained in a 4-membered ring.

In 1964, Belyaeva and Klyushina[64] showed that both ascorbic acid and unithiol reduced or nullified the 30–42% suppression of respiration caused by potassium dichromate in homogenates of liver, kidney, lungs and the stomach.

In 1971, Simavoryan[47] showed that i) dichromate injected subcutaneously into rats lowered the ascorbic acid content of their kidneys and ii) if he followed the dichromate injection within 30 minutes with a unithiol injection, it prevented to a great extent the lowering of the ascorbic acid concentration.

In our preliminary studies with unithiol we have found that at pH 7.4 (1 M Tris · HCl) and at 25 °C, the reaction between chromate and unithiol is first order in both chromate and unithiol with a second order rate constant of $26 \pm 1.6\ M^{-1}min^{-1}$.

7.2.3 2,3-Dimercaptosuccinic Acid

2,3-Dimercaptosuccinic acid (see Fig. 4 for structure) like unithiol has its two thiol groups separated by 2 carbon atoms. The result of our preliminary kinetic study indicate that its reaction with chromate at pH 7.4 (1 M Tris · HCl) and 25 °C, is first order in both chromate and unithiol and the second order rate constant is $3.8 \pm 0.2\ M^{-1}min^{-1}$.

7.2.4 Dithiothreitol (DTT)

DTT (see Fig. 4 for structure) has its two thiol groups separated by 4 carbon atoms, thus intramolecular oxidation would lead to a 6-membered ring and a bisthioester involving coordination of both thiol groups to Cr(VI) would involve a 7-membered ring.

The result of our kinetic studies indicate that at pH 7.4 (1 M Tris · HCl) and 25 °C, the reaction is first order in both chromate and DTT and the second order rate constant is $17.3 \pm 1.8\ M^{-1}min^{-1}$.

7.2.5 Lipoic Acid

Lipoic acid is an important dithiol in the context of this review, since it is a cellular metabolite which functions as a redox agent. Moreover, its two thiols are situated 3 carbon atoms apart and its intracellular oxidation is accompanied by ring closure. Its structure and redox reaction are illustrated in Eq. 21.

$$\begin{array}{l} CH_2{-}S \\ CH_2 \quad | \\ CH{-}S \\ (CH_2)_4 \\ COOH \end{array} + 2H^+ + 2e^- \rightleftharpoons \begin{array}{l} CH_2{-}SH \\ CH_2 \\ CH{-}SH \\ (CH_2)_4 \\ COOH \end{array} \qquad E_0' = -0.32\ V \tag{21}$$

The fact that 5 membered ring systems tend to be more stable than 4 membered rings would lead one to anticipate that the reduction of chromate by lipoic acid would be more facile than the above dithiols which would form 4- or 6-membered rings or react intermolecularly. However, the result of our preliminary kinetic study indicate a surprisingly slow rate of reaction with chromate at pH 7.4 (1 M Tris · HCl) and 25 °C. The reaction is first order in both chromate and lipoic acid and the second order rate constant was found to be $4.8 \pm 0.5\ M^{-1}min^{-1}$.

Thus the rate of chromate reduction for the various dithiols follows the order: unithiol > DTT > lipoic acid > 2,3-dimercaptosuccinic acid. If we compare the rates of the dithiols with the monothiols (see Table 2) the addition of a second thiol group does not confer great advantage with respect to the reduction of chromate under our standard conditions. Thus the transition state proposed by McAuley and Olatunji (see Fig. 5 and Ref. 38) for chromate-thiol reactions under acidic conditions, would not appear to be relevant under physiological conditions.

This result, together with the following observations:

i) the characteristic visible absorbance maxima of the thioester at about 420 nm, was not apparent during our reactions at pH 7.4 and

ii) our second order rate constant ($0.348\ M^{-1}s^{-1}$) and that of Hojo et al.[57] ($0.549\ M^{-1}s^{-1}$) in the reaction between chromate and penicillamine at pH 7.4 and 7 respectively is considerably different from that for k_2 ($0.143\ M^{-1}s^{-1}$) given by McAuley and Olatunji[38] for the same reaction carried out in the pH range 1–1.7 leads us to propose a change in the rate-determining step.

At acid pH, the rate-determining steps according to McCann and McAuley[37] (cysteine and chromate) and McAuley and Olatunji[38] (both glutathione and penicillamine with chromate) were those involved in the conversion of thioester to products either via reaction with a second thiol molecule or via reaction with a proton (see Sects. 4.1; 7.1.2., 7.1.3.).

We propose that at physiological pH, that the rate-determining step is the formation of the thioester from Cr(VI) and thiol. The reason for this change being twofold: firstly, the Cr(VI) species $HCrO_4^-$ is thought to be more labile to substitution than CrO_4^{2-} [37] and there will be much less of this species at pH 7.4 (about 2% of total Cr(VI), see Sect. 2.2.1) and secondly, under conditions where $HCrO_4^-$ is the dominant species, the rate of formation of the thioester is faster (about 20 times at pH 0) by a proton dependent route than by a proton independent route (see Table 4 in Ref. 51) for the reactions between chromate and cysteine, cysteamine and penicillamine. Moreover, the proton-independent rate constants for formation of the thioester in these reactions (Table 4 in Ref. 51) is of the same order of magnitude as our second order rate constants for these reactions at pH 7.4 (see Table 2).

In summary, the proposal that the rate-determining step in reactions between chromate and thiols at physiological pH is the rate of formation of thioester, explains why

i) there is no observable absorbance maxima at 420 nm (thioester) during the reaction, because the thioester will be consumed as soon as it is produced
ii) the addition of a second thiol group makes little difference to the overall rate of reduction, because the second thiol is involved after the rate determining step.

8 Other Low Molecular Weight Reductants

Other possible intracellular reductants include NADH, NADPH, ubiquinone, flavins, superoxide ion and ferrous ion. Control studies of microsomal reduction of chromate showed that NADH and NADPH reacted only very slowly with chromate at pH 7.4[65, 66]. Under acidic conditions (pH = 1–2) the kinetics of hexaquoiron(II) reduction of chromate was consistent with the rate law[67]:

$$\frac{-\,d[HCrO_4^-]}{dt} = \frac{K_1k_2[Fe^{2+}]^2[HCrO_4^-][H^+]^3}{[Fe^{3+}] + \frac{k_2}{k_{-1}}[Fe^{2+}][H^+]} \tag{22}$$

The value of K_1k_2 was $2.1 \times 10^8\ M^{-4}s^{-1}$ at 0 °C and ionic strength of 0.084 M[67]. The observed rate law was consistent with a mechanism whereby an equilibrium is established between chromium(VI)/chromium(V) and iron(II)/iron(III), and the rate determining step involves reaction of chromium(V) with iron(II).

$$Fe^{2+} + HCrO_4^- + 2\,H^+ \overset{K_1}{\rightleftharpoons} Fe^{3+} + H_3CrO_4 \tag{23}$$

$$Fe^{2+} + H_3CrO_4 + H^+ \xrightarrow{k_2} Fe^{3+} + Cr(IV) \tag{24}$$

When the reaction was run under conditions of high concentrations of iron(II) a reaction intermediate was observed which was postulated to be a $Fe(H_2CrO_4)^{2+}$ complex formed from reaction of Fe(II) with Cr(IV) or a $FeOCr^{4+}$ complex formed from reaction of Fe(II) with Cr(IV). Since the rate has a third-order dependence on hydrogen ion concentration, the reaction will be much slower under more basic conditions, e.g., pH 7.4, than under the acidic conditions used in this study.

9 Summary of the in vitro Reduction of Chromate by Low Molecular Weight Reductants

The leading low molecular weight candidates for the intracellular reduction of chromate (Table 2) are cysteine, ascorbate, cysteamine, glutathione and lipoic acid. The high intracellular levels[59)] of glutathione (0.8–8 mM) indicate that it may be the most effective of the four.

However, there is a very large and important question concerning the role of these reductants and the fate of chromate and its role as a carcinogen. Do these cellular reductants represent ultimate detoxifying agents for Cr(VI), producing kinetically inert, water soluble Cr(III) species, or do they, by producing activated Cr intermediates which can react with other cellular moities, enhance the damaging effects of chromium? Our preliminary results with the reaction between chromate and hemoglobin (see Sect. 10.1) indicate that the latter may be true. GSH increases the rate of the reaction between hemoglobin and Cr(VI) by a factor of about 3.

The results of chromate reduction by the model compounds listed in Table 2, allow us to predict the environment that the SH groups in a protein will require for them to react readily with chromate. The fastest reductants (Table 2) are either neutral (e.g., cysteine, penicillamine) or have a positive charge (cysteamine). However, charge is not the only factor affecting the kinetics, since the cysteine-chromate reaction is faster than the cysteamine-chromate reaction and the unithiol-chromate reaction is faster than the DTT-chromate reaction. The obvious advantage that cysteine has over cysteamine is its ability to chelate intermediate Cr species after the disulfide bond has been formed, as illustrated in Fig. 3. Such chelation would facilitate the change in coordination number required in going from Cr(VI) to Cr(IV) or Cr(III) (see Sect. 2).

More information is required before firmer conclusions can be drawn. Meanwhile as a starting hypothesis, the most reactive sites on protein towards chromate should have SH groups in a positively charged environment in which there are groups which can chelate to Cr species, and in which there are hydrogen ions readily available.

10 Purified Proteins

Not many purified proteins have been found to reduce chromate at physiological pH. Egg albumin and human plasma proteins were unreactive toward Cr(VI) at pH 6.4 and pH 7.35[68)]. Purified cytochrome c did not react with $^{51}CrO_4^{2-}$ [69)] and bovine serum albumin[70)], human serum albumin, serum globulin and soluble dermis protein did not react with $K_2Cr_2O_7$[50]. Cr(VI) was not reduced by fibrinogen, gelatin, casein or serum albumin after 48 h at pH 7 and 37 °C, however, gamma globulin and hemoglobin did cause some Cr(VI) reduction under the same conditions[40]. There was not a good correlation between the reducing ability of gamma globin and its cysteine and methionine content. In the process of developing a spectrophotometric assay for chromate reduction in systems containing NADPH, we found that no chromate reduction occurs with isocitrate dehydrogenase and NADPH.

10.1 Hemoglobin

While hemoglobin can exist in both the Fe(II) and Fe(III) state, it normally functions as an oxygen carrier in the Fe(II) state. However, each day about 1–2% of the hemoglobin is autooxidized to methemoglobin (Fe(III)) but in the red blood cell there are enzymes (reductases) which convert the oxidized hemoglobin to the Fe(II) state[71].

There are three general ways a low molecular weight redox agent might transfer electrons to the Fe atom in hemoglobin: a) it might attach to the Fe atom directly at the sixth axial position of the heme in an inner sphere mechanism; b) it might interact with the porphyrin system; and c) it might interact with an amino acid side chain on the polypeptide chain. In mechanism c) an electron moves either through some coupled amino acid system (such as the mechanism proposed by Winfield for cytochrome c[72]) or through the protein (8–10 Å) by an electron tunneling mechanism[73,74]. Huth et al.[75] have investigated the reduction of methemoglobin and metmyoglobin by Cr(II). They found second order rate constants for the reductions of 3.6×10^4 $M^{-1}s^{-1}$ and 5.9×10^4 $M^{-1}s^{-1}$ respectively for the reactions carried out at pH 6, 30.5 °C and ionic strength of 0.167. The kinetic products they obtained contained covalently bound chromium with Cr(III) to protein ratios of 1.0 for myoglobin and 4.0 for hemoglobin. These Cr(III)-protein complexes retained spectral properties and oxygen and CO binding properties identical to the untreated protein. They concluded that the Cr(III) was not bound at or near the sixth axial position and that the Cr(II) probably passed one electron to the Fe(III) via an exposed porphyrin edge. It was concluded that the chromium(III) was attached to amino acids 32–42 and/or 57–63 which contain close or adjacent aspartic acid and glutamic acid residues.

Although it has long been known that chromate labels the hemoglobin in the intact red blood cell, attempts to identify which type of hemoglobin and which chain was preferentially labelled have led to conflicting results. Intact red blood cells (RBC), hemolysates and stroma-free hemoglobin (Hb) preparations reacted with $^{51}CrO_4^{2-}$ and the ^{51}Cr remained bound to Hb[76]. Of the Hb's present in RBC, chromium preferentially labeled $Hb\text{-}A_3$[60,77–79] which is thought to be a breakdown product of Hb A, formed by conjugation with glutathione[80]. Several reports indicate that the ^{51}Cr preferentially labeled the β chains[77–79] which are known to contain a reactive sulfhydryl group[81]. However, a report by Chernoff indicated the α chains were preferentially labelled[82]. Prins reported the failure of chromatographically purified Hb-A and $Hb\text{-}A_3$ to react with $^{51}CrO_4^{2-}$[60]. Pearson reported that $^{51}CrO_4^{2-}$ reacted with approximately equal affinity with $Hb\text{-}A_1$, A_2 and A_3 purified by starch block electrophoresis and reacted with isolated β chains preferentially over α and γ chains[78]. Malcolm et al.[79] and Ebaugh et al.[77]

reported that electrophoretically purified Hb-A, and $Hb-A_3$ (oxy form) reacted to the same extent with $^{51}CrO_4^{2-}$ and resulted in a very low level of ^{51}Cr binding.

The fact that some reports indicate that the beta-chains are preferentially labelled warrants further consideration, since the beta-chains, unlike the alpha-chains, contain a SH group which readily reacts with the sulfhydryl reagents iodoacetamide and N-ethylmaleimide. Moreover, this SH group has been implicated in the oxidation of deoxyhemoglobin by Cu(II) and in reactions with SH groups of the low molecular weight reducing agents that reduce chromate at physiological pH (see Sect. 7). The beta chain SH group belongs to a cysteine residue which is number 93 in sequence. Its immediate environment in the beta chain (Fig. 6) clearly indicates that it occupies a very sensitive position being adjacent to histidine 92 whose side chain forms the fifth axial ligand to the Fe atom. Clearly, a very important question to ask is why nature should contrive to put a redox active group so close to a center whose normal function is not to change its oxidation state.

Winterbourn and Carrell[83] have suggested that the key role of the β cys 93 SH group is to protect the Hb from free radical attack, clearly a problem since it is known that some superoxide is generated by oxyhemoglobin[84]. The mechanism of how this might work is shown schematically in Fig. 7.

In considering to what extent the SH groups might play a role in the oxidation of hemoglobin by low molecular weight oxidants, we note that of the two oxidizing agents most studied in this respect, potassium ferricyanide and Cu(II), the evidence presented below indicates that the ferricyanide does not involve the SH groups while Cu(II) probably does. Allen and Jandl[85] showed that in the reaction with ferricyanide methemoglobin formation preceded the formation of disulfide bonds. Antonini et al.[86] confirmed this

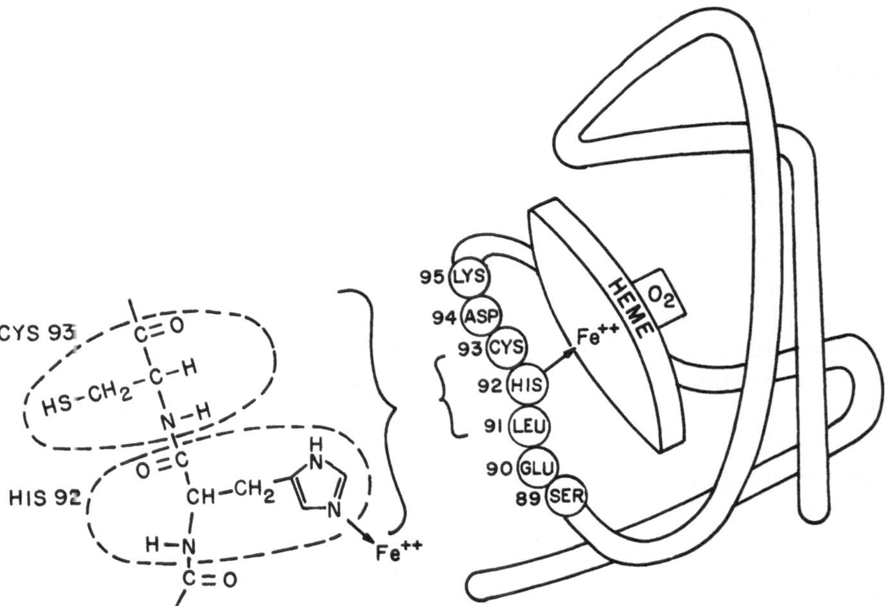

Fig. 6. Schematic illustration of the relative positions of cys 93, his 92 and the Fe atom in the beta chain of hemoglobin.

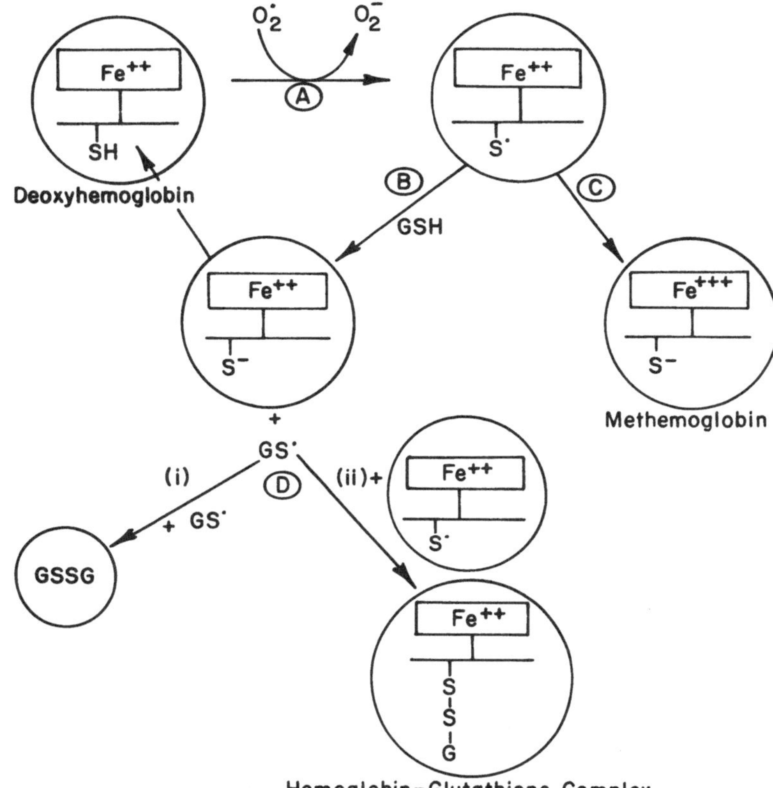

Fig. 7. Schematic representation of the possible role of the cys 93 SH group in scavenging for free radicals such as superoxide (after Winterbourn and Carrell[83]). (*A*) Interaction with superoxide or other free radical produces sulfur free radical at cysteine 93. (*B*) Interaction between this sulfur free radical and GSH produces GS˙ and deoxyhemoglobin. (*C*) Internal transfer of an electron from the Fe^{2+} to the S˙ might explain the origin of the 1–2% of methemoglobin formed in the RBC each day. (*D*) GS˙, formed in (*B*), has two fates: *i*) it usually will react with another GS˙ to form oxidized glutathione (GSSG) but *ii*) occasionally it might react with the S free radical at cys 93 and form the hemoglobin-glutathione complex associated with the aging of the RBC.

result by carrying out the ferricyanide oxidation in the presence of EDTA which is known to inhibit the interaction of ferricyanide and SH and found no diminution in the rate of ferricyanide oxidation. Moreover, the fact that ferricyanide can not only oxidize deoxymyoglobin, which has no cysteine residues at all, but does so at about 100 times the rate of the hemoglobin, strongly suggests that SH groups are not involved.

Brittain[87] has recently shown that Cu^{2+} oxidizes human deoxyhemoglobin to a half-met form in which only the beta chains are oxidized:

$$\underset{(\alpha_2^{2+}\beta_2^{2+})}{\mathrm{Hb}} \xrightarrow{2\,Cu^{2+}} \underset{(\alpha_2^{2+}\beta_2^{3+})}{\mathrm{Hb/Hb^+}}$$

The reaction went with both oxy and deoxyhemoglobin, the latter being faster (k = 196 $M^{-1}s^{-1}$ vs. k = 41 $M^{-1}s^{-1}$). The kinetics were interpreted as showing binding of Cu^{2+} ions to the protein as the slow step followed by rapid electron transfer.

Winterbourn and Carrell[83] had earlier shown that the Cu^{2+} bound at the β cys 93 SH s. They also indicated that the three forms of hemoglobin gave these interesting variations with Cu^{2+}:

a) deoxyhemoglobin is *not* converted to a disulfide, only the hemes are oxidized;
b) oxyhemoglobin forms a disulfide only after the hemes have been oxidized, and
c) methemoglobin forms a disulfide.

Their proposed mechanisms to explain these results are shown in Fig. 8.

Our studies with hemoglobin indicate that chromate oxidizes deoxyhemoglobin at pH 7 and that both alpha and beta chains are oxidized. The rate of Hb oxidation by chromate is very much slower than ferricyanide. Glutathione (GSH) increases the rate of Hb oxidation by chromate some two to threefold.

On the other hand, our studies with myoglobin indicate that there is little or no chromate oxidation of deoxymyoglobin since:

a) Incubating Cr(III) with deoxymyoglobin (20 h, room temperature) led to many Cr atoms bond per molecule of protein.
b) Incubating Cr(VI) with deoxymyoglobin in the presence of dithionite also produced a high level of binding.
c) But incubation with chromate and deoxymyoglobin led to very little, if any, Cr binding.

These results point towards SH involvement in the chromate oxidation of Hb. Moreover, the role of GSH in increasing the rate of chromate attack helps to explain some of the conflicting results listed at the beginning of this section since GSH would have been present in most of the Hb preparations that were from the RBC lysates.

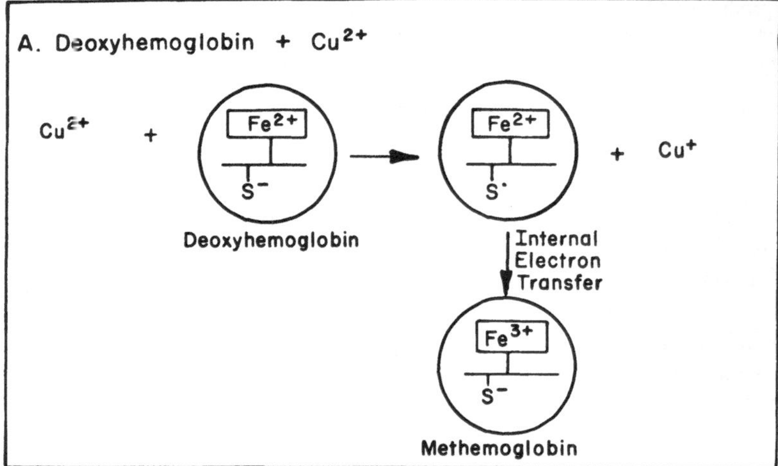

Fig. 8. Schematic illustration of the mechanisms proposed by Winterbourn and Carrell[83] to explain the different products obtained when Cu(II) reacts with **A** deoxyhemoglobin

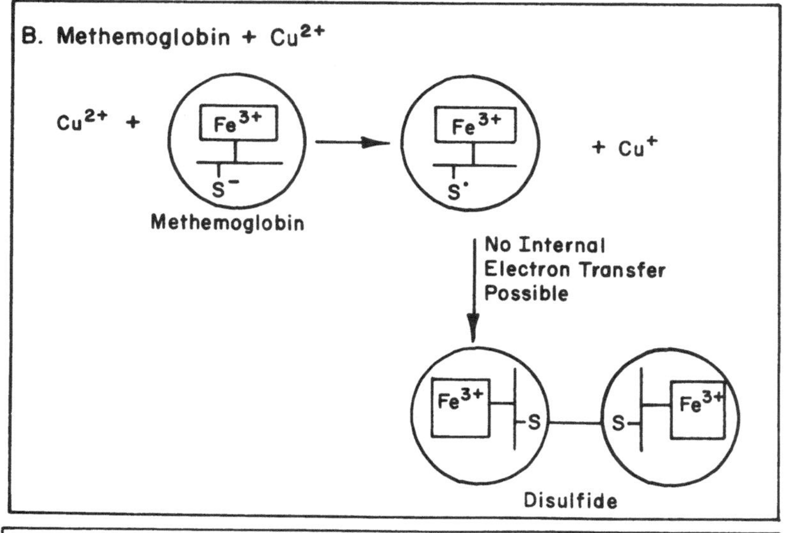

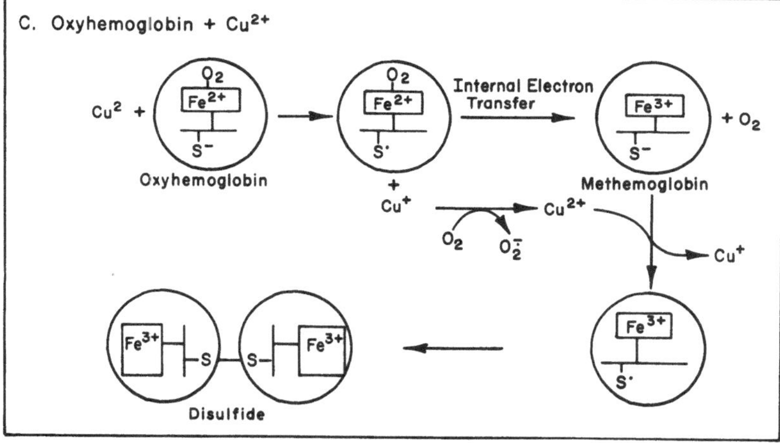

Fig. 8. Schematic illustration of the mechanisms proposed by Winterbourn and Carrell[83] to explain the different products obtained when Cu(II) reacts with **B** methemoglobin and **C** oxyhemoglobin

Fig. 8. **B** methemoglobin and **C** oxyhemoglobin

10.2 Glutathione Reductase

Chromate was found to inhibit human erythrocyte glutathione reductase (GSSG-R) activity in a NADPH-dependent manner[88, 89]. Incubation of erythrocytes with chromate had no effect on the activities of sixteen other enzymes including glyceraldehyde-3-phosphate dehydrogenase, hexokinase and enolase[88]. The inhibition of GSSG-R activity by chromate was accompanied by reduction of chromate to chromium(III). Chromium(III) did not inhibit GSSG-R activity[88]. The structure of human erythrocyte glutathione reductase has been determined[90]. In the oxidized form of the enzyme the active site was found to have a disulfide formed by Cys-46 and Cys-41. The mechanism of action was proposed to involve the formation of a thiolate ion at Cys-46 which is stabilized by a nearby protonated His in the stable reduced form of the enzyme. The reduced cys-41 was proposed to attack GSSG, release GSH and form a mixed disulfide with glutathione[29]. It is possible that the adjacent cys residues are involved in the reaction of GSSG-R with chromate.

10.3 Cytochrome c

Purified cytochrome c was not labeled after incubation with $^{51}CrO_4^{2-}$ in isotonic saline for 45 min at 37 °C[69]. Under identical conditions substantial labeling of cytochrome c occurred with $^{51}CrCl_3$. We have found that chromate oxidizes reduced cytochrome c very slowly at pH 7.5 (0.1 M Tris-acetate) and 25 °C.

Oxidized cytochrome c was reduced quickly and quantitatively by chromium(II) ion and a chromium(III)-bound ferrocytochrome c resulted[91]. The rate constant for the chromium(II) reduction of cytochrome c at 25 °C, pH 4.2 and ionic strength of 0.1 was 1.2×10^4 $M^{-1}s^{-1}$ [92], and was 3.1×10^3 $M^{-1}s^{-1}$ at 25 °C, pH 7.0 and ionic strength of 1.0[93]. Kinetic data was interpreted in terms of a mechanism of electron transfer from chromium(II) involving attack of Cr(II) adjacent to the Fe(III) center[93]. Analysis of the one-to-one chromium(III) : cytochrome c complex revealed that the chromium(III) cross-linked two peptide fragments located in the heme crevice by binding to tyrosine 67 and asparagine 52[94]. The chromium(III) bound to reduced cytochrome c did not affect the ability of the protein to be reöxidized with ferricyanide and then to be reduced with dithionite[91]. The chromium complex was oxidized by cytochrome oxidase at the same rate as the untreated ferrocytochrome c, however, the rate of reduction of the chromium complex by bovine heart submitochondrial particles was slower than that of untreated ferricytochrome c[91]. Thus, the binding of chromium(III) to cytochrome c appears to selectively inhibit its function in certain electron transfer reactions.

10.4 Plastocyanin

Although studies of chromate oxidation of reduced plastocyanin have not been reported, there have been extensive studies of the reduction of oxidized plastocyanin by chromium(II). The rate constant for the reduction of French bean plastocyanin by chromium(II) at 25 °C, pH 4.2 and ionic strength of 0.1 was 1×10^4 $M^{-1}s^{-1}$ [92]. Chromium(II) reduced French bean plastocyanin quantitatively and formed a stoichiometric complex with the protein[95]. Peptide analysis revealed that the chromium(III) was bound to a negative patch on the surface of the protein containing four carboxylate groups, Asp 42-Glu 43-Asp 44-Glu 45[95]. The chromium(III) binding site is approximately 12 Å away from the redox active copper center but is close to tyrosine 83. This residue is located in a hydrophobic channel which separates the copper center and the chromium(III) binding site. NMR studies with redox inactive analogs of inorganic electron transfer reagents identified the same binding site near the surface negative patch for the positively charged hexamminechromium(III) ion and tris(1,10-phenanthroline)chromium(III) ion[96]. Chromium(III) binding to French bean plastocyanin had no effect on photoreduction rates by chloroplasts, however, the rate of photooxidation by photosystem I reaction centers was significantly inhibited[97]. It was concluded that the bound chromium(III) interferred with the recognition site on plastocyanin used by the photosystem I reaction centers, whereas the bound chromium had no effect on the electron transfer from cytochrome f through His 87 to the copper center of plastocyanin. As was observed in the case of cytochrome c, bound chromium(III) also selectively inhibits some electron transfer reactions of plastocyanin.

11 Organelles

In addition to the soluble reductants in the cytoplasm, the various cellular organelles possess possible chromate-reductase activity. Both the endoplasmic reticulum (ER) and mitochondria (mt) possess electron-transport chains whose components are potentially capable of reducing chromate. Since the endoplasmic reticulum is contiguous with the nuclear membrane the presence of chromate-reductase activity in the ER implies a similar activity in the nuclear membrane.

11.1 Endoplasmic Reticulum

We have identified a NAD(P)H-dependent chromate reductase activity in rat liver microsomes[65, 66]. Kinetic studies using microsomes from phenobarbital-pretreated rats and specific inhibitors of microsomal enzymes led to the identification of cytochrome P-450 as the chromate reductase[66].

Chromate was reduced to chromium(III) by rat liver microsomes and NADPH in vitro[65, 66]. Reduction of hexavalent chromium to trivalent chromium required the presence of both microsomal protein and NADPH cofactor. Heat denaturation of microsomes resulted in the loss of their ability to reduce chromate. Essentially no chromate reduction was observed by the cofactor in the absence of microsomal proteins. We have found that NADH also served as a cofactor for the microsomal reduction of chromate. However, at equivalent concentrations of NADPH and NADH, the rate of reduction of limiting amounts of the substrate chromate was slower with NADH than with NADPH cofactor.

The rate of chromate reduction using NADH as a cofactor did not vary with the source of microsomes. The same kinetic parameters were found with control, phenobarbital-induced and 3-methylcholanthrene-induced microsomes. Therefore there was no correlation of rate of chromate reduction with the activity of NADH-Cytochrome P-450 reductase in the microsomes which was ~ 1.25 times greater in the control set than in either the phenobarbital- or 3-methylcholanthrene-induced microsomes. The cytochrome b_5 content of the microsomes, however, did not vary ($\pm 5\%$) with the source of microsomes.

In contrast, using NADPH as cofactor phenobarbital-induced microsomes metabolized chromate with an apparent second order rate constant ~ 1.25 times greater than that found with control or 3-methylcholanthrene-induced microsomes. The increased rate of chromate reduction seen with phenobarbital-induced microsomes correlated with the greater NADPH-Cytochrome P-450 reductase activity of these microsomes compared with control and 3-methylcholanthrene-induced microsomes. The differences in rate could also be ascribed to the greater content of cytochrome P-450 induced by phenobarbital compared to control. 3-Methylcholanthrene-induced microsomes had a greater content of Cytochrome P-448 compared to control and our results, therefore, suggest that Cytochrome P-448 may not be significantly involved in chromate reduction. It has been shown that at least four different forms of Cytochrome P-450 are present in phenobarbital-induced rat liver microsomes which exhibit different substrate specificities and spectral characteristics[98, 99]. One or more of these forms may be involved in the reduction of chromate to chromium(III).

The NADPH- and NADH-Cytochrome P-450 reductase activities of microsomes have been shown to be inhibited by 3-pyridinealdehyde-NAD[102]. The rate of reduction of chromate by microsomes using NADH cofactor was significantly decreased by the presence of 3-pyridinealdehyde-NAD; with NADPH cofactor the rate was substantially slower at equivalent concentrations of inhibitor and cofactor.

NADPH-Cytochrome P-450 reductase has been shown to be specifically inhibited by 2′-AMP[101, 102]. Studies with varying concentrations of the 2′-AMP inhibitor and varying concentrations of the substrate chromate showed that 2′-AMP appeared to be a competitive inhibitor of chromate reduction by microsomes and NADPH or NADH. The involvement of NADPH- and NADH-Cytochrome P-450 reductase in the microsomal metabolism of chromate was, therefore, confirmed.

Metyrapone has been shown to inhibit Cytochrome P-450 activity and bind to the reduced and oxidized form of the enzyme[103]. Increasing concentrations of metyrapone progressively decreased the rate of chromate reduction by microsomes and NADPH. Metabolism by both control and phenobarbital-induced microsomes was affected similarly by metyrapone which appeared to be a competitive inhibitor of chromate reduction.

Carbon monoxide has been shown to bind the reduced form of Cytochrome P-450[104] and inhibit its enzymatic activity[105, 106]. The rate of reduction of chromate by both control and phenobarbital-induced microsomes and NADPH was dramatically decreased by carbon monoxide. The rate progressively decreased with increasing concentration of carbon monoxide in the system. These studied implicate the involvement of Cytochrome P-450 in the in vitro metabolism of chromate by microsomes.

Chromium(V) was detected in incubation mixtures of chromate with microsomes and NADPH[108]. EPR signals characteristic of Cr(V) appeared within 20 s after initiation of the reaction and persisted for 80 min. The rapid formation of Cr(V) in this system implies that a direct one-electron transfer from the microsomal-electron-transport cytochrome P-450 system to chromate is a likely step in the mechanism of reduction.

Purified pig liver NADPH-cytochrome P-450 reductase had a low level of chromate-reductase acitivity. This enzyme ($E_0' = -320$ mV, -190 mV), like the NADPH cofactor (-320 mV), although thermodynamically capable of reducing chromate must have a kinetic barrier to the reaction. Upon adding purified PB-induced rat liver cytochrome P-450 to the purified reductase the NADPH-dependent chromate reductase activity was greatly increased. Thus, reconstitution studies with the purified enzymes confirm our previous studies with microsomes which implicated cytochrome P-450 as the chromate reductase. Cytochrome P-450 appears to be able to transfer electrons to chromate in a facile manner.

Addition of chromium(VI) to homogenates containing the microsomal fraction of rat lung tissue dramatically inhibited the benzo[*a*]pyrene (BP) hydroxylase activity known to be associated with the microsomal cytochrome P-450 system[109]. Chromium(III) had no effect on the BP hydroxylase activity of the rat lung homogenates[109], however, chromium(III) inhibited the BP hydroxylase activity of rat liver microsomes[110]. Chromate also inhibited the BP hydroxylase activity of mouse liver homogenates[111]. Thus in addition to being metabolized by the microsomal cytochrome P-450 system, chromate as well as the product of its metabolism, chromium(III), can inhibit the normal activities of the system.

11.2 Mitochondria

Calcium chromate has been shown to induce cytoplasmic *petite* mutations in mitochondria of *Saccharomyces cerevisiae*[112]. Calcium chromate also dramatically depressed the content of the mitochondrial gene products cytochrome aa_3 and cytochrome b, in whole yeast cells[112]. Chromate (~8 nM) was readily taken up by rat thymocytes and after 30 min 9% of the ^{51}Cr was found in the mitochondria although 62% was found in the nuclei[113]. Isolated rat thymus mitochondria and nuclei readily took up $^{51}CrO_4^{2-}$[113]. After one hour incubation of Erlich ascites tumor cells with $^{51}CrO_4^{2-}$ (380 μM), 17% of the chromium was distributed in the nuclear fraction and 12% was in the mitochondrial-microsomal fraction[114]. Levels of chromium in rat liver mitochondria reached a plateau six hours after i.v. injection of chromate (0.02 mg/kg) and remained at that level through 5 days[115]. Liver nuclear chromium levels in the same animals, although similar to mitochondrial levels at 6 h, reached a maximum at 12 h and steadily decreased after that time. Therefore the nuclear chromium levels were lower than the mitochondrial chromium levels at later times (24–120 h) after injection. The subcellular distribution of chromium in the liver of rats injected i.v. with chromate (0.56 mg/kg) was also found to be time dependent in another study[116]. The distribution of chromium in rat liver mitochondria increased from 5% at 15 min to 21% at 72 h and also increased in the nuclear fraction from 22% at 15 min to 52% at 72 h. Incubation of isolated rat liver mitochondria with chromate (0.3–16.6 μM) for 30 min resulted in uptake of 86–90% of the chromium into the mitochondria in a firmly bound (non-dialysable) form[116]. Langard[116] suggested that the chromate was reduced by the electron transport chain of the mitochondrial iner membrane.

Our studies revealed that freshly isolated rat liver mitochondria are capable of reducing chromate. Heat denaturation of the mitochondria destroyed their chromate reductase activity. Addition of the substrates malate and glutamate dramatically enhanced the rate of chromate reduction by mitochondria. Succinate, also, had an effect on the rate of chromate reduction. Malate and glutamate were not capable of reducing chromate in the absence of mitochondria. The purified enzymes malate dehydrogenase and glutamate dehydrogenase had no chromate reductase activity under any conditions. These results indicate that there is a malate/glutamate-dependent chromate reductase activity in mitochondria. This activity may reside early in the electron transport chain or may be associated with the cytochrome P-450-like system of the matrix[117].

12 Conclusion

The studies reviewed above indicate that there are many possible sites for chromate reduction in the cell. The studies with small molecules suggest that only ascorbate and thiol-containing molecules are efficient at reducing chromate under physiological conditions. Thus, in the cell cytoplasm the significant reactive species are expected to be glutathione, cysteine and ascorbate. Very few soluble proteins react with chromate. There are two soluble proteins in erythrocytes, hemoglobin and glutathione reductase, which are capable of reducing chromate. In microsomes two enzymes of the electron transport system, NADPH-cytochrome P-450 reductase and cytochrome P-450, metabolize chromate. Unidentified enzymes in mitochondria also are capable of reducing chromate. Thus, of the enzymes identified as chromate-reductases two are heme proteins involved in oxygen metabolism and two are NADPH-dependent flavoproteins. However, not all heme proteins are chromate-reductases as evidenced by the inactivity of cytochrome c and myoglobin. The NAD(P)H-dependent enzymes which are not flavoenzymes, isocitrate dehydrogenase, glutamate dehydrogenase and malate dehydrogenase, do not possess chromate-reductase activity. More studies are required to determine whether these are general phenomena related to chromate metabolism.

Chromate reduction within cells is not expected to be random and nonselective, rather only certain small molecules and enzymes appear capable of reducing chromate at a significant rate under physiological conditions. The selective metabolism of the carcinogen chromate by cellular constituents may lead to selective damage of cell components by chromium metabolites and alteration of their normal functions.

Postscript. Our continuing kinetic studies of thiol-chromate reactions (see Sect. 7) indicate that the logarithms of the observed rate constants (1 M Tris · HCl, pH 7.4, 25 °C) show an inverse linear relationship to the microscopic pK_a's of the thiols.

Acknowledgements. The investigations described from the author's laboratory were supported by grant BC-320 from the American Cancer Society and by an A. P. Sloan Research Fellowship.

13 References

1. Norseth, T.: Environ. Health Perspect. *40*, 121 (1981)
2. Sirover, M. A.: ibid. *40*, 163 (1981)
3. Zakour, R. A., Kunkel, T. A., Loeb, L. A.: ibid. *40*, 197 (1981)
4. Wetterhahn Jennette, K.: ibid. *40*, 233 (1981)
5. Jennette, K. W.: Biol. Trace Element Res. *1*, 55 (1979)
6. Warren, E. et al.: Mutat. Res. *90*, 111 (1981)
7. Tsapakos, M. J., Hampton, T. H., Wetterhahn Jennette, K.: J. Biol. Chem. *256*, 3623 (1981)
8. Tsapakos, M. J., Wetterhahn, K. E.: submitted
9. Espenson, J. H.: Acc. Chem. Res. *3*, 347 (1970)
10. "C. R. C. Handbook of Chemistry and Physics", 53rd ed. (West, R. C. ed.), D. 111, Cleveland CRC press, 1972
11. Baes, C. F., Mesmer, R. E.: The Hydrolysis of Cations, pp. 211–219, New York, John Wiley, 1976
12. Cotton, F. A., Wilkinson, G.: Advanced Inorganic Chemistry, 4th ed., p. 733, New York, John Wiley, 1980
13. Taube, H., Gordon, G.: Inorg. Chem. *1*, 69 (1962)
14. Kon, H.: J. Inorg. Nuc. Chem. *25*, 933 (1963)
15. Mahapatro, S. N., Krumpolc, M., Rocek, J.: J. Am. Chem. Soc. *102*, 3799 (1980)

16. Espenson, J. H.: ibid. *86*, 5101 (1964)
17. Greenblatt, M., Banks, E., Post, B.: Acta Crystallogr. *23*, 166 (1967)
18. Russev, p., Mitewa, M., Bontchev, P. R.: J. Inorg. Nuc. Chem. *43*, 35 (1981)
19. Krumpolc, M., Rocek, J.: J. Am. Chem. Soc. *98*, 872 (1976)
20. Hasan, F., Rocek, J.: ibid. *94*, 9073 (1972)
21. Hasan, F., Rocek, J.: ibid. *94*, 8964 (1972)
22. Srinivasan, V., Rocek, J.: ibid. *96*, 127 (1974)
23. Hasan, F., Rocek, J.: J. Org. Chem. *39*, 2612 (1974)
24. Hasan, F., Rocek, J.: J. Am. Chem. Soc. *97*, 1444 (1975)
25. Krumpolc, M., De Boer, B. G., Rocek, J.: ibid. *100*, 145 (1978)
26. Hintze, R. E., Rocek, J.: ibid. *99*, 132 (1977)
27. Beattie, J. K., Haight, G. P., Jr.: Progr. Inorg. Chem. *17*, 93 (1972)
28. Huss, E., Klemm, W.: Angew. Chem. *66*, 468 (1954)
29. Griffith, W. P.: Coord. Chem. Rev. *5*, 459 (1954)
30. Taube, H.: Electron Transfer Reactions of complex Ions in Solution, p. 18, New York, Academic Press, 1970
31. Beck, M. T., Bardi, I.: Acta Chim. Hung. *29*, 283 (1961)
32. Cooper, J. N. et al.: Inorg. Chem. *12*, 2075 (1973)
33. Espenson, J. H., King, E. L.: J. Am. Chem. Soc. *85*, 3328
34. Espenson, J. H.: ibid. *86*, 5101 (1964)
35. Sullivan, J. C.: ibid. *87*, 1495 (1965)
36. Birk, J. P.: ibid. *91*, 3189 (1969)
37. McCann, J. P., McAuley, A.: J. Chem. Soc. Dalton 783 (1975)
38. McAuley, A., Olatunji, M. A.: Can. J. Chem. *55*, 3335 (1977)
39. Samitz, M. H., Pomerantz, H.: Arch. Ind. Health *18*, 473 (1958)
40. Samitz, M. H., Katz, S.: J. Invest. Derm. *43*, 35 (1964)
41. de Meester, P. et al.: Inorg. Chem. *16*, 1494 (1977)
42. Agent, C. L., Pennington, D. E.: Am. Chem. Soc. Abst. 34th. SW Regional Meet., Corpus Christi, TX, Inorg. 23 (1978)
43. McAuliffe, C. A., Murray, S. G.: Inorg. Chim. Acta Rev. 103 (1972)
44. Rajka, G., Vincze, E., Csanyi, G.: Borgyogy es venerol. *31*, 18 (1955)
45. Samitz, M. H., Shrager, J., Katz, S.: Indust. Med. Surg. *31*, 427 (1962)
46. Samitz, M. H., Katz, S.: Arch. Env. Health *11*, 770 (1965)
47. Simarvoryan, P. S.: Tr. Erevan Med. Inst. *15*, 219 (1971)
48. Samitz, M. H., Scheiner, D. M., Katz, S.: Arch. Env. Health *17*, 44 (1968)
49. Baldea, I., Munteano, L.: Stud. Univ. Babes-Bolyai *25*, 24 (1980)
50. Mali, J. W. H., Van Kooten, W. J., Van Neer, F. C. J.: J. Invest. Derm *41*, 111 (1963)
51. McAuley, A., Olatunji, M. A.: Can. J. Chem. *55*, 3328 (1977)
52. Olatunji, M. A., McAuley, A.: J. Chem. Soc. Dalton 682 (1975)
53. Muirhead, K. A., Haight, G. P., Jr.: Inorg. Chem. *12*, 1116 (1973)
54. Baldea, I., Niac, G.: ibid. *6*, 1232 (1968)
55. Pearson, R. G.: J. Chem. Ed. *45*, 581 (1968)
56. Sugiura, Y., Hojo, Y., Tanaka, H.: Chem. Pharm. Bull *20*, 1362 (1972)
57. Hojo, Y., Sugiura, Y., Tanaka, H.: J. Inorg. Nuc. Chem. *39*, 1859 (1977)
58. de Meester, P., Hodgson, D. J.: J. Chem. Soc. Dalton 1604 (1977)
59. Rabenstein, D. L., Guevemont, R., Evans, C. A.: Metal Ions in Biol. Systems *9*, 103 (1979)
60. Prins, H. K.: Vox Sang 7, 370 (1962)
61. Oehme, F. W.: Clin. Toxicol. *5*, 215 (1972)
62. Braun, H. A., Lusky, L. M., Calvery, H. O.: J. Pharm. Exp. Therap. (Supp.) *87*, 119 (1946)
63. Cole, H. N.: Arch. Dermat. Syph. *67*, 30 (1953)
64. Belyaeva, L. N., Klyushina, L. V.: Vopr. Gigieny Prof. Patol. I Toksikol., Sb., Sverdlovsk 413 (1964)
65. Gruber, J. E., Jennette, K. W.: Biochem. Biophys. Res. Commun. *82*, 700 (1978)
66. Garcia, J. D., Jennette, K. W.: J. Inorg. Biochem. *14*, 281 (1981)
67. Espenson, J. H.: J. Am. Chem. Soc. *92*, 1880 (1970)
68. Grogan, C. H., Oppenheimer, H.: Arch. Biochem. Biophys. *56*, 204 (1955)
69. Ingrand, J.: Biochem. Pharmacol. *15*, 1649 (1966)

70. Baetjer, A. M. et al.: Arch. Ind. Health *12*, 258 (1955)
71. Rifkind, J. M.: Biochemistry *13*, 2475 (1974)
72. Winfield, M. E.: J. Mol. Biol. *12*, 600 (1965)
73. Hopfield, J. J.: Biophys. J. *18*, 311 (1977)
74. Potasek, M. J., Hopfield, J. J.: Proc. Natl. Acad. Sci. USA *74*, 3817 (1977)
75. Huth, S. W. et al.: J. Am. Chem. Soc. *98*, 8467 (1976)
76. Gray, S. J., Sterling, K.: J. Clin. Invest. *29*, 1604 (1950)
77. Ebaugh, F. G. et al.: Proc. IX Congress Int. Soc. Hematol. *3*, 489 (1962)
78. Pearson, H. A.: Blood *22*, 218 (1963)
79. Malcom, D., Ranney, H. M., Jacobs, A. S.: ibid. *21*, 8 (1963)
80. Jonxis, J. H. P.: J. Pediat. *59*, 765 (1961)
81. Green, D. W., Ingam, U. M., Perutz, M. F.: Proc. Roy. Soc. *A 225*, 287 (1954)
82. Chernoff, A. I.: Nature *192*, 327 (1961)
83. Winterbourn, C., Carrell, R. W.: Biochem. J. *165*, 141 (1977)
84. Misra,H. P., Fridovich, I.: J. Biol. Chem. *247*, 6960 (1972)
85. Allen, D. W., Jandl, J. H.: J. Clin. Invest. *40*, 454 (1961)
86. Antonini, E., Brunori, M.: J. Biol. Chem. *244*, 3909 (1969)
87. Brittain, T.: Biochem. J. *187*, 803 (1980)
88. Koutras, G. A. et al.: J. Clin. Invest. *43*, 323 (1964)
89. Koutras, G. A. et al.: Brit. J. Haemat. *11*, 360 (1965)
90. Schulz, G. E. et al.: Nature *273*, 120 (1978)
91. Kowalsky, A.: J. Biol. Chem. *244*, 6619 (1969)
92. Dawson, J. W. et al.: Proc. Natl. Acad. Sci. USA *69*, 30 (1972)
93. Yandell, J. K., Fay, D. P., Sutin, N.: J. Am. Chem. Soc. *95*, 1131 (1973)
94. Grimes, C. J., Piszkiewicz, D., Fleischer,E. B.: Proc. Natl. Acad. Sci. USA *71*, 1408 (1974)
95. Farver, O., Pecht, I.: ibid. *78*, 4190 (1981)
96. Cookson, D. J., Hayes, M. T., Wright, P. E.: Biochim. Biophys. Acta *591*, 162 (1980)
97. Farver, O., Shahak, Y., Pecht, I.: Biochemistry *21*, 1885 (1982)
98. Werringloer, J., Estabrook, R. W.: Arch. Biochem. Biophys. *167*, 270 (1975)
99. Thomas, P. E. et al.: J. Biol. Chem. *251*, 1385 (1976)
100. Strittmatter, P.: ibid. *234*, 2665 (1959)
101. Neufield, E. F., Kaplan, N. O., Colowick, S. P.: Biochim. Biophys. Acta *17*, 525 (1955)
102. Phillips, A. H., Langdon, R. G.: J. Biol. Chem. *237*, 2652 (1962)
103. Hildebrant, A. G.: in: Biological Hydroxylation Mechanisms (Eds. Boyd, G. S., Smellie, R. M. S.) pp. 79–102, New York, Academic Press, 1972
104. Omura,T., Sato, R.: J. Biol. Chem. *239*, 2370 (1964)
105. Omura, T. et al.: Fed. Proc. *24*, 1181 (1965)
106. Conney, A. H. et al.: J. Biol. Chem. *243*, 3912 (1968)
107. Gillette, J. R.: Metabolism *20*, 215 (1971)
108. Wetterhahn Jennette, K.: J. Am. Chem. Soc. *104*, 874 (1982)
109. Dixon, J. R. et al.: Cancer Res. *30*, 1068 (1970)
110. Thomson, R., Webster, I., Kilroe-Smith, T. A.: Environ. Res. *7*, 149 (1974)
111. Calop, J., Burckhart, M.-F., Fontanges, R.: Eur. J. Toxicol. *9*, 271 (1976)
112. Egilsson, U., Evans, I. H., Wilkie, D.: Molec. gen. Genet. *174*, 39 (1979)
113. Scaife, J. F., Vittorio, P. V.: Can. J. Biochem. *42*, 503 (1964)
114. Rajam, P. C., Jackson, A.-L.: Proc. Soc. Exp. Biol. Med. *99*, 210 (1958)
115. Sawato, Y. et al.: J. Pharm. Dyn. *3*, 17 (1980)
116. Langard, S.: Biol. Trace Element Res. *1*, 45 (1979)
117. Niranjan, B. G., Avadhani, N. G.: J. Biol. Chem. *255*, 6575 (1980)
118. Mahler, H. R., Cordes, E. H.: Biological Chemistry, 2nd Ed., p. 30, New York, Harper and Row, 1971
119. Lehninger, A. L.: Biochemistry, 2nd Ed., p. 479, New York, Worth Publishers, Inc., 1975
120. Clark, W. M.: Oxidation-Reduction Potentials of Organic Systems, pp. 502–505, Huntington, Robert E. Krieger Publishing Co., 1972
121. Beutler, E., Duron, O., Kelly, B. M.: J. Lab. Clin. Med. *61*, 882 (1963)

Mineralization in Biological Systems

Stephen Mann

Inorganic Chemistry Laboratory, South Parks Road, Oxford, OX1 3QR, U.K.

The major solid state principles involved in mineralization in biological systems are discussed. Three major biological control factors of mineralization are described; structural, spatial, and chemical control. Factors determining nucleation on organic surfaces, mineral growth, mineral structure, and morphology are reported. Oriented growth of minerals on organic matrices can occur by three processes; (i) lattice matching (epitaxis) between the organic matrix and depositing crystal faces, (ii) surface structural relationships between the organic matrix and depositing crystal faces, (iii) ordered aggregation of preformed mineral particles. Several *a priori* reasons indicate that epitaxis is unlikely to be a major process of oriented growth in biomineralization. The presence and importance of biogenic amorphous minerals is also described.

1 Introduction

The number of different minerals known to be formed in biological systems has increased enormously in recent years (Table 1.1). A recent review[1)] showed that the majority of these biogenic minerals are calcium-containing, often hydrated and may be monocrystalline, polycrystalline, or amorphous and can be of different functional value to the organism (Table 1.2). The precision of replication of these minerals in terms of structure, morphology and distribution in each species suggests that there are highly specific mechanisms of control of crystal nucleation, growth and spatial organisation within the biological system.

Two fundamental processes have been suggested in the formation of biogenic minerals[1)]. Firstly, there is the process of *"biologically-induced"* deposition, in which the mineral precipitates as a result of interactions between the biological activity of the organism and the surrounding physical environment. In this process mineral deposition is under minimal biological control. Secondly, under strict genetic control, there is the *biologically-controlled* process of "organic matrix-mediated" deposition in which the induction of nucleation and growth of minerals is controlled by the presence of an organic matrix generated by cellular activity. Thus many biominerals, unlike their inorganic counterparts, are composite materials often consisting of mineral sub-units ordered within an organic matrix ("mosaic biominerals") (Fig. 1.1). A special case is where the mosaic biomineral has monocrystalline properties (as determined, for example, by X-ray diffraction); in this case each sub-unit is crystalline and oriented in a preferential direction within the organic matrix (referred to here as "iso-oriented mosaic crystal") (Fig. 1.1).

Although it is clear that the mediation of the mineralization reaction by organic constituents is of major importance, the role of chemical and spatial factors in regulating biomineralization must not be overlooked. It has often been inferred that the main process of biological regulation of mineralization by organic matrices is a crystallochemical-controlled process, with the biominerals being *structurally* defined by the atomic arrangement of the organic framework. However, there may be several other more likely functions of the organic matrix. The aim of this paper is to discuss the solid state principles involved in mineral deposition under conditions of chemical, spatial, and structural

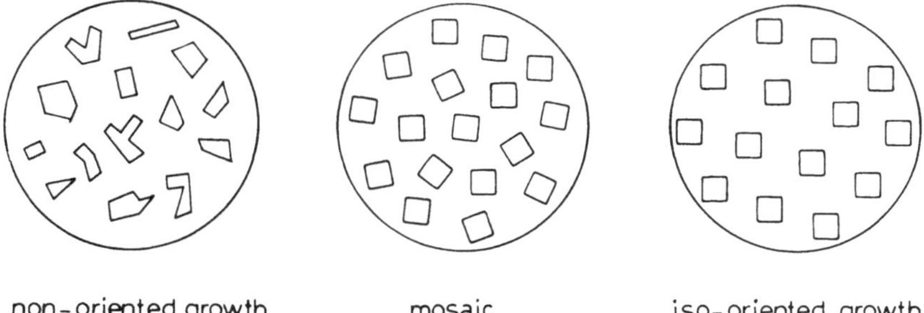

Fig. 1.1 a–c. Growth of inorganic solids on organic substrates; **a**, non-oriented growth; **b**, mosaic structure from oriented growth; **c**, iso-oriented growth. In **b** and **c** the orientation of mineral sub-units is determined by the spacial design of the organic matrix

Table 1.1. Diversity and phylum distribution of biomineralization products in extant organisms[1)]

Kingdom	Monera	Protoctista														Fungi		Animalia											Plantae	
Phylum		Dinoflagellata	Haptophyta	Bacillariophyta	Phaeophyta	Rhodophyta	Chlorophyta	Zygnematophyta	Rhizopodea	Siphonophyta	Charophyta	Heliozoata	Radiolariata	Foraminifera	Mixomycota	Ciliophora	Basidiomycota	Deuteromycota	Porifera	Coelenterata	Platyhelminthes	Ectoprocta	Brachiopoda	Annelida	Mollusca	Arthropoda	Sipuncula	Echinodermata	Chordata	Bryophyta	Trachaephyta
Carbonates:																															
Calcite	+	+	+			+				+	+			+	+				+	+	+	+	+	+	+	+	+	+	+	+	+
Aragonite	+		?		+	+	+			+				+					+	+		+		+	+	+	+		+		+
Vaterite						+																			+	+			+		+
Monohydrocalcite	+																								+				+		
Amorph. hydr. carb.															+						+				+	+			+		
Phosphates:																															
Dahllite	+															+		+											+		?
Francolite																							+		+				+		
$Ca_3Mg_3(PO_4)_4$																								+							
Brushite																									+						
Amorph. dahllite precursor																					+			+	+			+			
Amorph. brushite precursor																									+						
Amorph. whitlockite precursor																					+			+	+						
Amorph. hydr. ferric phosphate																								+	+			+			
Halides:																															
Fluorite																									+	+		+			
Amorph. fluorite precursor																									+				+		

Oxalates:																					
Whewellite			?	+		+					+	?					+			+	+
Weddelite											+	?				+		+	+		
Sulfates:																					
Gypsum														+						?	+
Celestite									+												
Barite					+		+														
Silica:																					
Opal		+			?			+	+	+			+		+	+	+	+			+
Fe-Oxides:																					
Magnetite	+															+	+		+		
Maghemite	?																				
Goethite																+					
Lepidocrocite													+			+					
Ferrihydrite	+										+	+			+	+			+	+	+
Amorph. ferrihydrates										+					+				+		
Mn-Oxides:																					
"Todorokite"	+																				
Fe-Sulfides:																					
Pyrite	+																				
Hydrotroilite	+																				

Table 1.2. The function of the main inorganic solids formed in biological systems[2)]

Cation	Anion	Formula	Function
Calcium	Carbonate	$CaCO_3$	Exoskeleton Gravity device Eye lens
	Phosphate	$Ca_{10}(PO_4)_6(OH)_2$	Skeletal Ca store Piezo-electric
	Oxalate	$Ca(COO)_2 \cdot 2\,H_2O$	Deterrent Cytoskeleton Ca store (plants)
	Sulphate	$CaSO_4 \cdot 2\,H_2O$	Gravity device Ca, S store (plants)
Iron	Oxide	Fe_3O_4 α-FeOOH γ-FeOOH $5\,Fe_2O_3 \cdot 9\,H_2O$	Magnetic device Teeth Teeth Fe store
Silicon	Oxide	SiO_2	Skeletal Deterrent (plants)
Magnesium	Carbonate	$MgCO_3$	Skeletal

control in biological systems, and to discuss the relative importance of the possible functions of the organic matrix in these processes.

Three main questions are discussed;

(i) What factors influence the nucleation of biominerals?
(ii) What factors influence biomineral growth?
(iii) What factors determine the final mineral structure, orientation, morphology and size?

2 Mineralization in Biological Systems

Figure 2.1 shows a general scheme for biological control in mineralizing systems to be developed in this paper. There is a hierarchy of control mechanisms inherent in biomineralization ranging from mineralization under minimal biological control (biologically-induced mineralization) to precise regulation of the chemistry of the mineralizing zone, the spatial organisation of the mineral particles, and the structural properties of the solid phase. The mechanisms chosen in a particular biological system may often be determined by the complexity of the organism. Thus biomineral formation can be considered, from a wider perspective, to be determined by the evolutionary status of the organism.

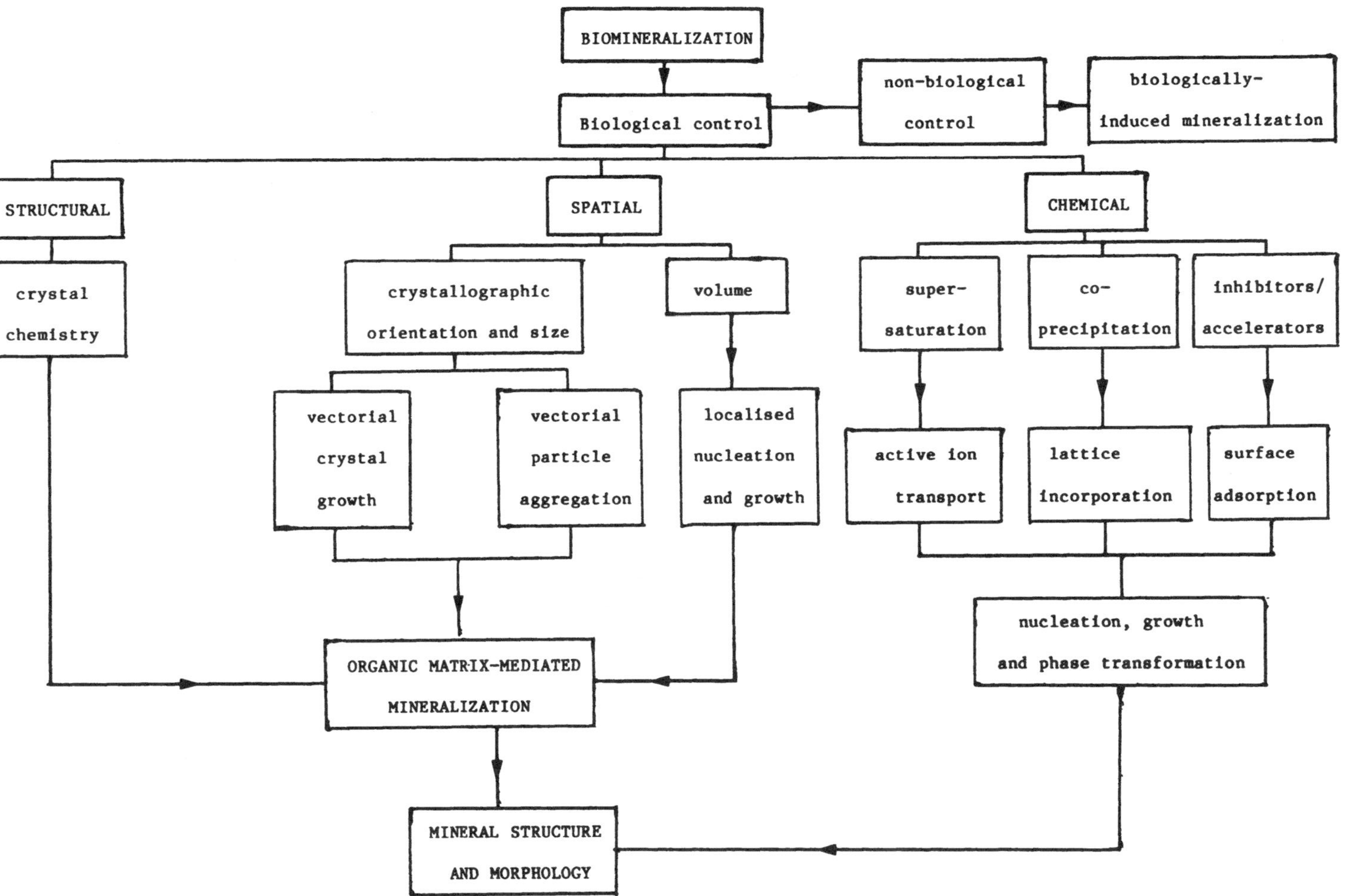

Fig. 2.1. General scheme for biological control in mineralizing systems

2.1 Biologically-Induced Mineralization

In this process mineralization takes place as a result of interactions between the biological activity of the organism and the surrounding physical environment. Such processes are under minimal biological control and are therefore not discussed in detail in this paper. They are characterised by bulk extracellular and/or intercellular mineralization without the elaboration of organic matrices. For example, calcification in some green algae[3,4], takes place in intercellular spaces through the interaction of biological activity and the ions in the external medium. Since the water in which these algae live is generally supersaturated with respect to $CaCO_3$ the reduction in CO_2 concentration in the intercellular spaces by photosynthesis results in the precipitation of a polycrystalline deposit of $CaCO_3$, as seen from the following chemical equilibrium;

$$Ca^{2+}_{(aq)} + 2\,HCO_{3\,(aq)}^{-} = CaCO_{3(s)} + CO_{2(g)} + H_2O_{(aq)}$$

2.2 Biologically-Controlled Mineralization

The rationalization presented in this paper describes the biological control of intra- and extracellular mineralization in terms of three fundamental processes; structural, spatial and chemical control. These factors can be incorporated into a generalised scheme for biomineralization (Fig. 2.2). Mineralization will occur within localised volumes, such as within the micro-compartments of lipid vesicles, or at major extracellular sites often some distance away from the initial stages of the mineralization process. In systems where biomineralization is located within confined volumes of small dimension, i.e. there is no major (bulk) mineralizing front, the vesicles are likely to be the major sites of precipitation. Membrane-bound vesicles can be derived from the Golgi apparatus of the cell[5–7] and can function as compartments for localised deposition or as accumulation and transport centres for aqueous ions[5,6,8,9]. Mineralization can take place within vesicles of pre-determined shape and size (vesicle dimensions may change during the process of mineralization) by biological control over the intravesicular ion concentrations. Control over ion transport can be through the selective permeability of the vesicle membrane which may be controlled by cellular processes (genetically and environmentally induced) or by intravesicular processes via chemical feed-back loops (information transfer). The ions to be precipitated would then be transported into the vesicles where, depending on the chemical properties of the internal compartment and the nature of the vesicle surface, they could precipitate at sites on the surface of the inner membrane.

There is growing evidence that cellular vesicles are also important in biological systems in which mineralization occurs at a major extracellular site. In these cases the role of such vesicles is likely to be in the chemical regulation of this relatively large biological volume. Thus a bulk oriented mineralizing zone could be developed by transport of the vesicles followed by the precipitation of the released intravesicular ions at the extracellular matrix by processes of oriented growth. Alternatively, vesicles containing initial mineral phases could be transported to the extracellular site where the bulk mineral phase could be developed by the release of these particles followed by spatial alignment within the extracellular organic matrix.

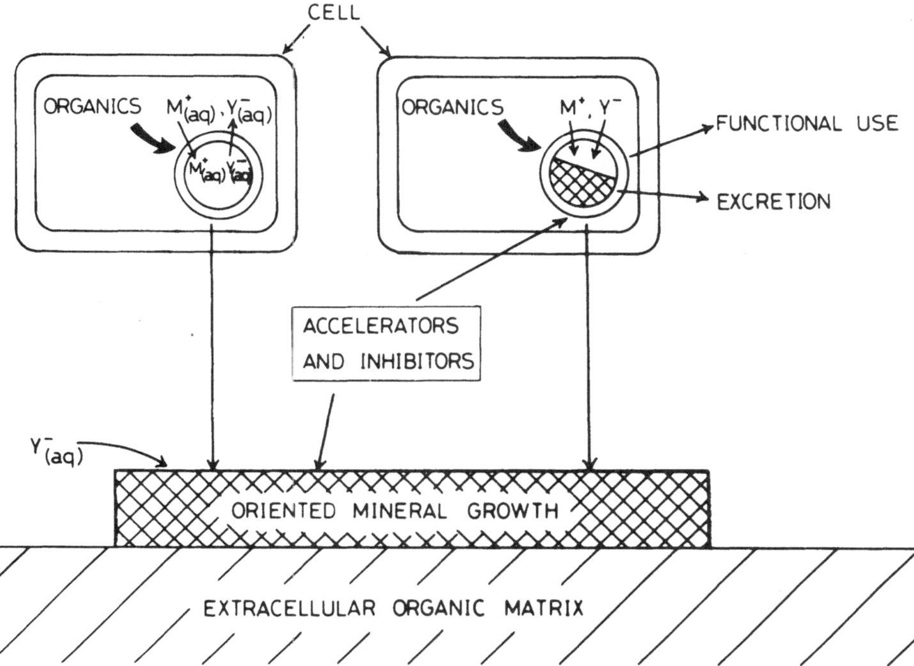

Fig. 2.2. Generalised scheme for biomineralization. Mineralization may occur on an extracellular matrix or within intracellular vesicles. Intravesicular deposits can be later transferred to an extracellular site where they can be aligned by aggregation on an organic framework. Alternatively, vesicles can act as accumulation centres for aqueous ions which are then transported to an extracellular mineralizing site where they are released and precipitated. In this scheme the mineral front may be oriented and mineral growth regulated by accelerators and inhibitors

3 Chemical Control of Biomineralization

A fundamental process of biologically-controlled mineralization will be through the chemical control of ions and molecules at the mineralizing site. In general, nucleation and growth can be controlled solely through the levels of supersaturation and co-precipitating ions resulting in specific structural and morphological mineral states. Control in this manner will be bio-energetically determined, ion concentrations at the mineralizing site being set by the rate of ion movement (active transport) across membrane boundaries.

A further mechanism of chemical control will be through the interaction of mineral phases with inorganic and organic molecules functioning as accelerators or inhibitors of mineral development. Such interactions may then lead to specific structural and morphological mineral states. Specificity is thus determined through accelerator/inhibitor molecular design, accelerator/inhibitor concentration levels, and temporal control over molecular transport into the mineralizing zone. No organic matrix is necessarily required in processes of mineralization under chemical control since mineral structure, morphology and size can, in principle, be determined solely by the physico-chemical properties of the mineralizing environment.

3.1 Theories of Nucleation

There are two primary forms of nucleation; homogeneous and heterogeneous nucleation. Homogeneous nucleation occurs due to the spontaneous formation of nuclei in the bulk of supersaturated solutions whereas heterogeneous nucleation involves the formation of nuclei on the surface of a substrate present in the aqueous medium. Despite the fact that homogeneous nucleation is a rarely observed phenomenon the principles of this process form the necessary background for the understanding of nucleation at the surface of an organic substrate.

If the concentration of a pure solution is increased above the activity product of its constituent ions the new solid phase will not occur until a certain degree of supersaturation has been achieved. A pre-requisite for the phase transformation is the formation of stable embryonic clusters of ions or molecules through statistical fluctuations in the supersaturated solution. Prior to nucleation there is continuous formation and dissolution of these clusters since the growth of the aggregates proceeds against a gradient of free energy required to create the new interface. Only when the expenditure of this energy is overcome by the energy released in the formation of bonds in the bulk solid phase is a critical stable nucleus attained. The free energy of formation of any nucleus, ΔG_N, is then given by,

$$\Delta G_N = \Delta G_{surface} + \Delta G_{bulk} \quad (1)$$

where ΔG_{bulk} is a negative quantity and $\Delta G_{surface}$ refers to the solid-liquid interface, and can be written for a spherical nucleus, radius r, as

$$\Delta G_N = 4\pi r^2 \sigma - \frac{4\pi r^3 \Delta G_v}{3 V_m} \quad (2)$$

where σ is the interfacial free energy per unit surface area, ΔG_v the free energy change per mole associated with the liquid-solid phase change, and V_m the molar volume. A plot of ΔG_N against r is shown in Fig. 3.1. The maximum value of the free energy ΔG_{N^*}, at a critical nucleus size r*, corresponds to the energy of activation required for the phase transformation and its dependence on the degree of supersaturation can be calculated from classical nucleation theory[10] as,

$$\Delta G_{N^*} = \frac{16\pi\sigma^3 v^2}{3\,(kT \log_e S)^2} \quad (3)$$

where S is the supersaturation defined as c/c_0 where c is the concentration of ions in the solution and c_0 the ionic concentration at equilibrium; ν, the volume per molecule; k, the Boltzmann constant; and T, the temperature. It can be shown by differentiation of Eq. 2 that

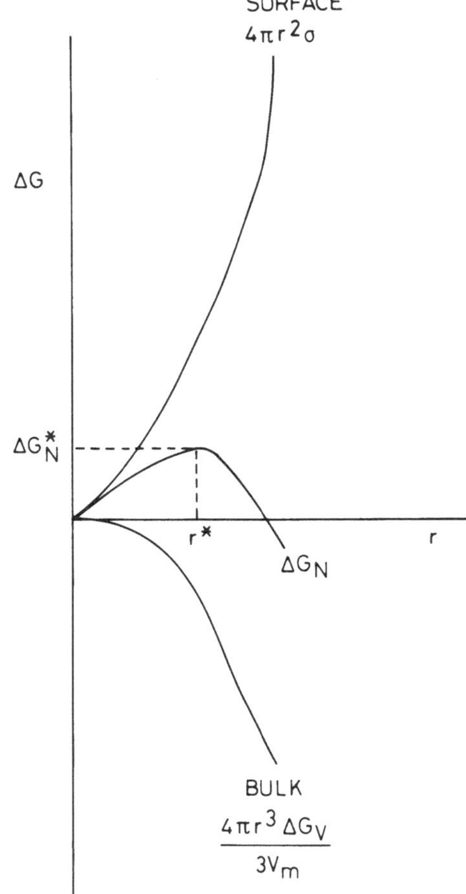

Fig. 3.1. Plot of free energy of nucleation, ΔG_N, against cluster size, r. The free energy of nucleation is given by the difference between surface and bulk energy terms. Only above a critical size, r*, corresponding to ΔG_{N^*}, does nucleation become energetically favourable

$$r^* = \frac{2\,\sigma V_m}{\Delta G_v} \quad (4)$$

indicating that, for constant V_m and ΔG_v, smaller critical nuclei are favoured for clusters of lower interfacial energy.

The presence of an organic substrate, such as a protein/carbohydrate surface or the inner membrane surface of a vesicle, within the precipitating medium will modify ΔG_{N^*} such that

$$\Delta G_{N^{**}} \leqslant \Delta G_{N^*} \quad (5)$$

where $\Delta G_{N^{**}}$ is the free energy associated with the formation of a stable nucleus through heterogeneous nucleation. The exact mechanism of this process is unclear, it is generally considered to be the result of a local ordering process arising from interactions across the interface. Thus most organic substrates will aid nucleation.

This thermodynamic approach illustrates in terms of a limiting supersaturation the energy requirements involved in nucleation. Since the majority of precipitation reactions in biological systems are likely to occur under kinetic conditions, the rate of nucleation will be of upmost importance. The rate of homogeneous nucleation J_N, can be considered as the rate at which nuclei surmount the maximum in the free energy curve of Fig. 3.1 and can be expressed as,

$$J_N = A \exp(-\Delta G_{N^*}/kT) \qquad (6)$$

Measurable nucleation rates have been reported between $10^{6\,\pm 3}\ m^{-3}\ s^{-1}$ and the pre-exponential factor is of the order $10^{36\pm 3}\ m^{-3}\ s^{-1}$ [11]. Since ΔG_{N^*} is related to the supersaturation as given in Eq. (3) there is a rapid increase in J_N as the supersaturation reaches a critical value (S*) (Fig. 3.2).

From the rate equation it can be deduced that a lowering in ΔG_{N^*} to $\Delta G_{N^{**}}$ by nucleation at a substrate increases J_N by modification of the exponential term in this expression. Thus, heterogeneous nucleation will be favoured at lower supersaturation levels than for those required for homogeneous nucleation (Fig. 3.3).

The structure of the critical nucleus will determine to a large degree the structure of the initial mineral phase. However, there is little structural information about the initial states formed in nucleation since the embryonic clusters are of multi-atomic dimensions and often exist for only transient moments. The precipitation of ionic crystals on organic substrates will involve stable nuclei of differing structure depending on the strength of the interactions between the precipitating ions, and the ions and solvated water molecules in the critical nucleus. Two extreme possibilities exist (Fig. 3.4). Firstly, the initiating nucleus resembles a piece of the bulk crystalline phase. This model, treated through the classical formulation of nucleation theory involves strong interactions between ions in the

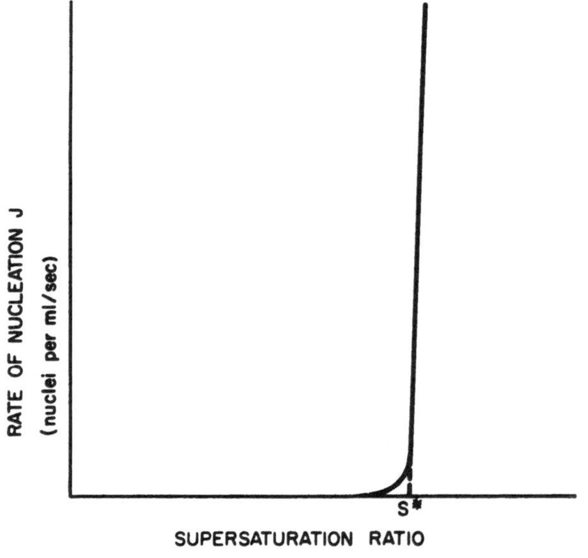

Fig. 3.2. The relationship between the rate of homogeneous nucleation, J_N, and the degree of supersaturation. There is a rapid rise in J_N at a critical value S*

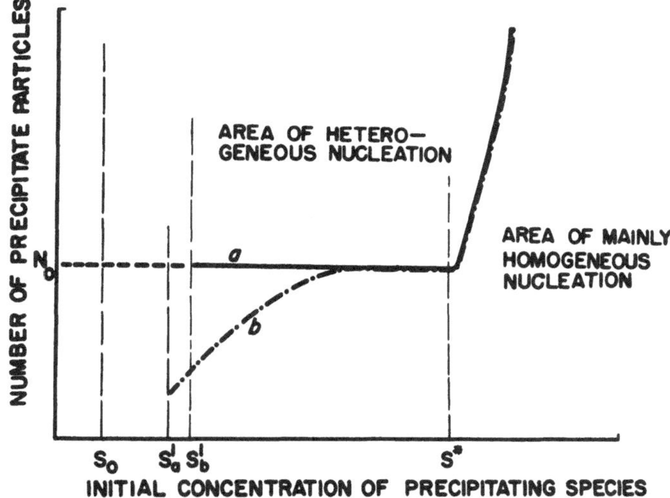

Fig. 3.3 a, b. Typical curves for the number of precipitated particles produced as a function of the initial precipitant concentration; **a**, in the presence of N_0 impurity particles of equal nucleation efficiency; **b**, in the presence of N_0 impurities of varying nucleation efficiency. Homogeneous nucleation occurs at a much higher level of supersaturation, S*, than for heterogeneous nucleation, S'; S_0 is the solubility of the precipitant[12)]

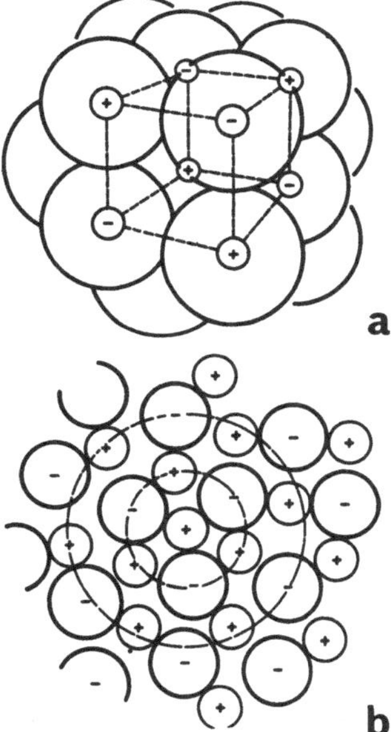

Fig. 3.4 a, b. Possible two-dimensional structures for critical ionic nuclei, **a**, the nucleus resembles a micro-volume of the bulk crystalline phase; **b**, the nucleus contains weakly bonded ions in a diffuse, highly solvated structure[12)]

nuclei so that the lattice solvation is overcome and the configuration becomes that of a small piece of bulk crystal at the substrate surface. The ions in the critical nuclei can be considered to be relaxed to some degree from their normal unit cell positions but there will be a definite correspondence between the normal and initial lattice parameters of the depositing phase. The second model of nuclei structure considers stable clusters to be composed of weakly interacting hydrated ions forming a diffuse highly solvated solid phase at the substrate surface. In this case there is no correspondence between cluster geometry and the lattice configuration of the final crystalline mineral. In both models of nuclei structure the number of ions involved in the critical nucleus is of the order 10–1000, i.e. 5–20 Å in size.

In general, then, the nucleation activation energy and rate will be determined by the surface energy of the critical nucleus and by the level of supersaturation within the system. Both these factors can be biologically influenced through the molecular design of organic substrates and the regulation of ion concentration gradients.

3.2 Theories of Crystal Growth

In order to understand the chemical factors which can determine the morphology of biogenic solids some knowledge of crystal growth processes in inorganic systems is discussed Three related processes are considered;

(i) crystal growth from pure solution in the absence of intermediate phases.
(ii) crystal growth from impure solution with minimal crystallographic modification.
(iii) crystal growth from structurally-modified precursers.

3.2.1 Crystal Growth from Pure Solution (no Intermediates)

The theoretical growth of a pure crystal phase was first given by Kossel[13] and Stranski[14] and later by Becker and Doring[15, 16]. These theories considered crystal growth to occur through surface nucleation of further ions or molecules at the crystal surface in a similar manner to the process of primary heterogeneous nucleation. The surface of the growing crystal contains sites of high binding energy at which further deposition can occur. These sites are steps or kinks in the surface, kink sites having the greater binding energy since they have three faces in contact with the crystal surface (Fig. 3.5)

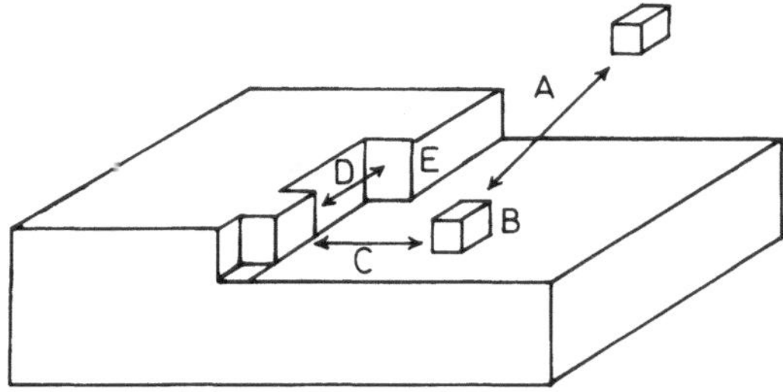

Fig. 3.5 A–E. Diagrammatical representation of a crystal surface showing the process of growth from pure solution. **A**, ions or molecules diffuse to the surface from bulk solution; **B**, dehydration and adsorption at the surface; **C**, two-dimensional diffusion across the surface to active sites; **D**, one-dimensional diffusion along a step to a kink or dislocation site; **E**, incorporation into the active site. Each step may be energetically demanding

Ions or molecules are adsorbed onto the crystal surface and diffuse to active sites such that the crystal grows by step-wise addition of ions to the surface. The model predicts that an energy barrier has to be overcome once a surface layer is completed due to the requirement for a new surface nucleation site. The most rapid crystal growth should then occur on those crystal faces which are the most heavily covered in kinks. However, it appears that only a small number of kinks persist at any given time[17)]. Thus theories of crystal morphology being dependent on the surface energies of different faces inducing different rates of growth in different directions[18)] have mainly fallen into disuse. However, the elimination of the high energy (110) plane in crystals of cubic structure is often considered to be due to the high growth rate on this plane (Fig. 3.6).

For surface nucleation to occur, supersaturation values of 25–50% are usually required. Thus this model of crystal growth does not explain the observation that many crystals grow at supersaturations as low as 1%. The screw-dislocation theory of crystal growth[19, 20)] explains this observation by considering real crystals to contain faces intersected by dislocation ledges which are sites of further crystal growth (Fig. 3.7). Growth takes place by the spreading of this ledge over the surface. Since the ledge is self-perpetuating the need for surface nucleation is obviated and the ledge winds itself up into a closed spiral as it advances during growth.

The kinetics of growth processes will be dependent on the slowest step in the following series of events (Fig. 3.5); (i) bulk transport of ions or molecules to the crystal surface (diffusion), (ii) dehydration of ions at the surface, (iii) adsorption of ions or molecules at the surface, (iv) two-dimensional diffusion across the surface to active sites, (v) one-dimensional diffusion along the step to the kink or dislocation site, (vi) incorporation into the kink or dislocation site. Each of these steps may be energetically demanding. The activation energy for bulk diffusion is about 6 kT whilst for ion hydration it is of the order 15–25 kT[11)]. However, the rates of different mechanisms may be comparable such that the rate control can alter from one mechanism to another upon a change of ionic concentration or particle size. Thus, for consecutive mechanisms such as

transport $\rightarrow$ surface adsorption $\rightarrow$ integration

the slowest step is rate determining whilst for parallel rate determining mechanisms such as

screw dislocation steps/surface nucleated steps

the faster mechanism is decisive.

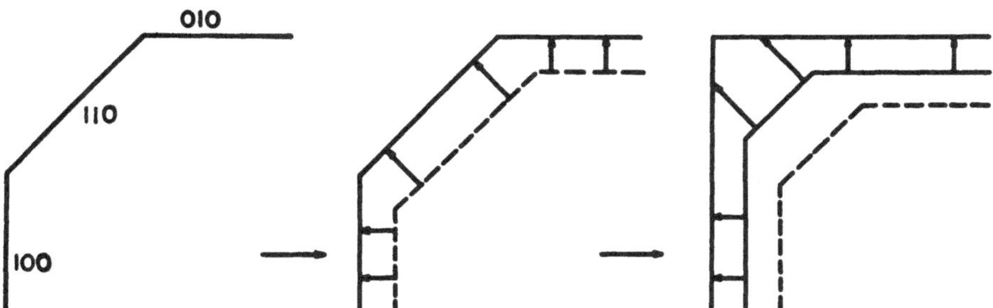

Fig. 3.6. Schematic diagram of the elimination of a high energy (110) crystal face. Growth occurs most rapidly on the high energy face such that this face becomes eventually eliminated

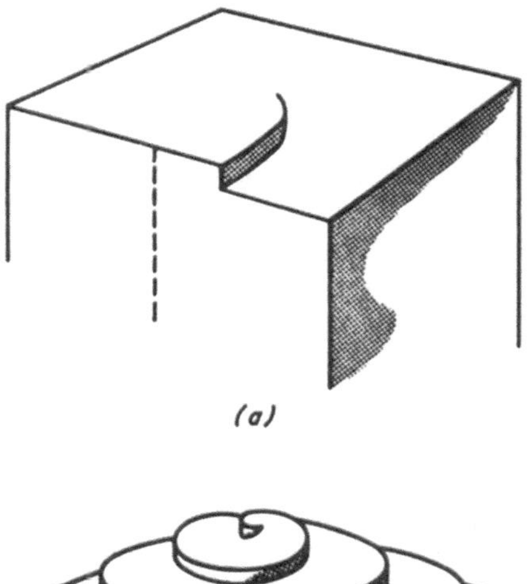

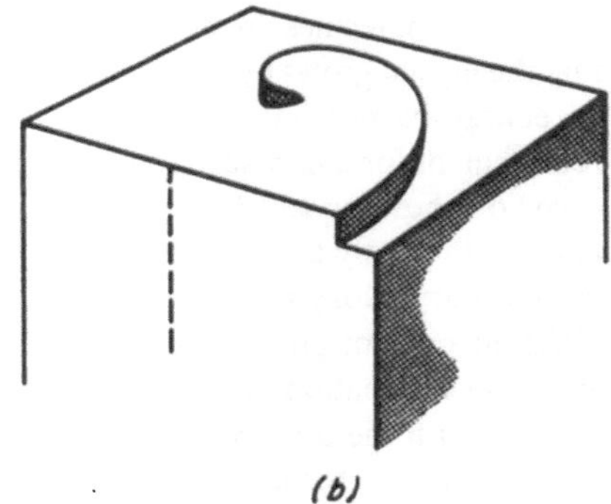

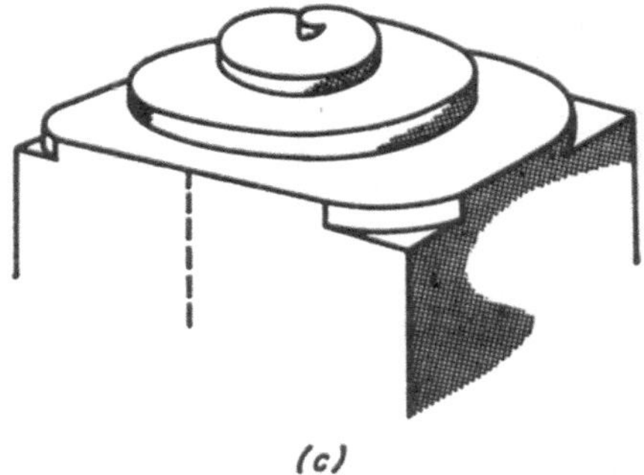

Fig. 3.7 a–c. Diagram showing the growth of a crystal at a screw dislocation. Deposition of new material at the screw dislocation edge occurs at a uniform rate but the anchor dislocation centre causes the relative rate of "winding" to be larger at the origin. Consequently, the screw dislocation winds into a growth spiral[12)]

First-order kinetics are predicted at high supersaturation since bulk diffusion is likely to be the limiting factor under these conditions. At lower supersaturation, theory predicts a second-order dependence on supersaturation for screw-dislocation growth[21)]. However, observed kinetics are often complicated, being dependent on the concentration and nature of the ions at the growing surface[22, 23)]. The presence of ion-pairs rather than hydrated ions may be important in the surface growth steps[24)].

3.2.2 *Crystal Growth from Impure Solution*

Solution components can be incorporated into a host lattice such that they cause modifications in crystal growth, morphology and chemical properties. Isomorphic replacement of host ions by ions of similar size and charge results in minimal structural modification but may lead to chemical differences within different regions of the crystal since the composition of the solid phase may be heterogeneous, being dependent on the relative concentrations of precipitating ions at any given time. It has been reported that the replacement of OH by F in the hydroxyapatite lattice results in an increase in unit cell size and therefore in the size of individual crystallites, thus reducing their solubility[25, 26)]. Alternatively, ions of different charge may be coprecipitated resulting in modifications of the host lattice, crystal growth and crystal chemistry. For example, Fig. 3.8 illustrates the

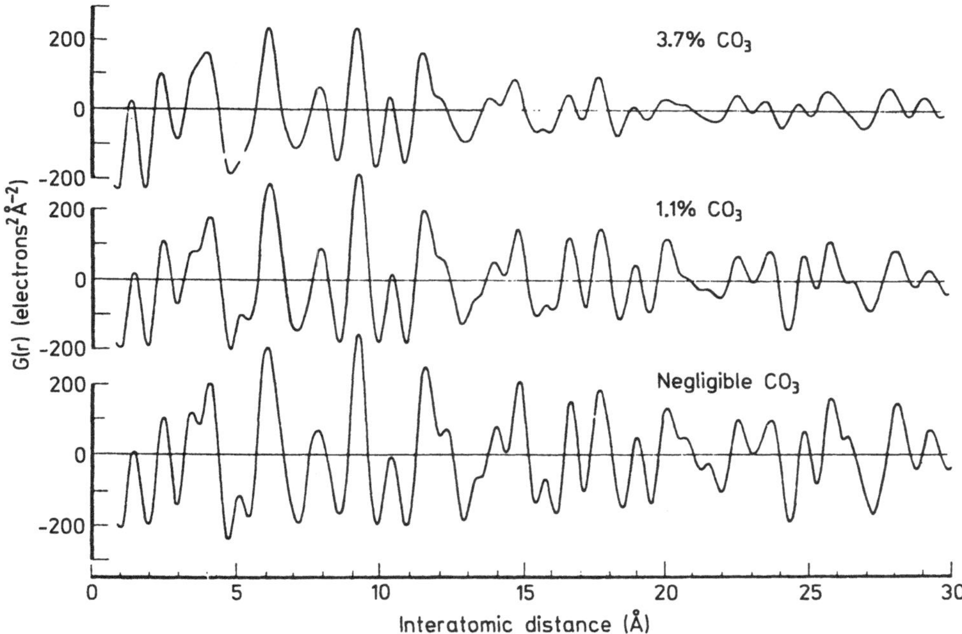

Fig. 3.8. Reduced radial distribution functions, G(r), of three synthetic hydroxyapatites containing 0%, 1.1%, 3.7% CO_3, respectively. The inclusion of CO_3 in the apatite structure results in distortion of the lattice, observed as the diminished intensity of G(r) with increasing r and the decreased resolution of certain G(r) peaks[27)]

effect of CO_3 substitution for PO_4 on the regularity of the interatomic distances in hydroxyapatite crystals.

The miscibility of coprecipitating ions will depend on the relative ion sizes, charge, and polarization; the degree of ion-size mismatch being usually not greater than 15%[28)]. Different ions will be favoured by different host lattices. For example, Mg^{2+} favours coprecipitation with the calcite lattice whereas Sr^{2+} is selectively incorporated in the aragonite lattice due to the isomorphic relationships between $MgCO_3$ and calcite, and $SrCO_3$ and aragonite[29, 30)]. Two extreme possibilities exist in coprecipitation. Firstly, the solid phase is perfectly homogeneous with host guest ions in thermodynamic equilibrium. This condition will be obtained under conditions of slow crystal growth since the guest ions have to diffuse through the host lattice presumably through Schottky and Frenkel defects. However, under kinetic conditions it is more probable that localised distribution of guest ions will occur. Localised distribution may also be favoured over homogeneous mixing where the latter condition would result in large distortions in the host lattice. In this case, the presence of localised zones could lower the energy of the crystal in the same way that aggregation lowers the energy of a colloidal system.

The presence of coprecipitating ions may markedly affect the rate of crystal growth through inhibition of active surface sites in crystal formation. This phenomenon may be highly selective, being dependent on the crystallographic structure of the host lattice. Thus, dissolved Mg^{2+} has no effect on the rate of crystal growth of aragonite although it severely retards the formation of calcite[29]. Mg^{2+} ions are readily adsorbed onto the calcite surface and incorporated into its crystal structure. It has been suggested that inhibition of calcitic growth is due to the non-equilibrium incorporation of Mg^{2+} into the growing calcite crystals which causes them to be considerably more soluble than pure calcite[29].

3.2.3 Crystal Growth from Structurally-Modified Precursers

The growth of crystals from precurser phases of different crystallographic structure involves a process of ion-translocation to new lattice coordinates. Phase transformations can occur via surface dissolution of the precurser followed by reprecipitation of a second phase often upon particles of the initially formed solid. Alternatively, the second phase can be formed via an *in situ* solid state transformation particularly when there is a close structural match and low interfacial energies between the two phases. A much studied example is the conversion of amorphous calcium phosphate (ACP) to hydroxyapatite (HAP) in aqueous solution[31–33]. The main processes involved appear to be diffusion or surface-reaction controlled and are very dependent on pH and temperature. An important observation has been that both ACP and crystalline phases occur together during the period of transformation indicating that the original amorphous surface provided the primary sites for crystal nucleation[34]. The amorphous particles are enveloped by the crystalline mass through growth of the initial crystals and by creation of new crystals from the original ones suggesting that the conversion process may be autocatalytic in nature[31]. Crystal growth is away from the interior of the amorphous particles which may indicate a process of solution translocation of Ca^{2+} and phosphate ions from the ACP phase rather than an *in situ* solid state process.

The first intermediate phase formed at pH 7.4 is octacalcium phosphate (OCP)[32]. The transformation of OCP to HAP is thermodynamically favoured and occurs partly as an *in situ* solid state process, possibly due to the close structural arrangements and low interfacial energies between the two phases[35]. The process appears to take place by hydrolysis of the OCP phase; one unit cell thickness of OCP (d_{100} = 18.68 Å) hydrolysing to form a two unit cell thickness of HAP (2 d_{100} = 16.32 Å) and resulting in a contraction of 2.36 Å along the [100] axis[36].

In summary, the factors that affect crystal growth in inorganic solutions are; (i) level of supersaturation, (ii) mechanistic processes (screw-dislocations, surface nucleation, surface reactions, bulk diffusion, and dehydration), (iii) concentration of other (co-precipitating) ions and molecules in the system, (iv) pH and temperature.

3.3 Chemical Control of Biomineral Structure

Biomineral structure will be determined by the nucleation and growth characteristics of the mineralizing system and therefore can be biologically regulated by the control of the chemistry of the mineralizing zone. In general, it is probable that the chemical control of nucleation proceeds cooperatively with other mechanisms of nucleation control such as the controlled precipitation of biominerals on organic substrates (see Sect. 4). The possible influence of chemical factors on the modulation of nucleation and growth can be discussed with reference to the energy states and energy pathways involved in these

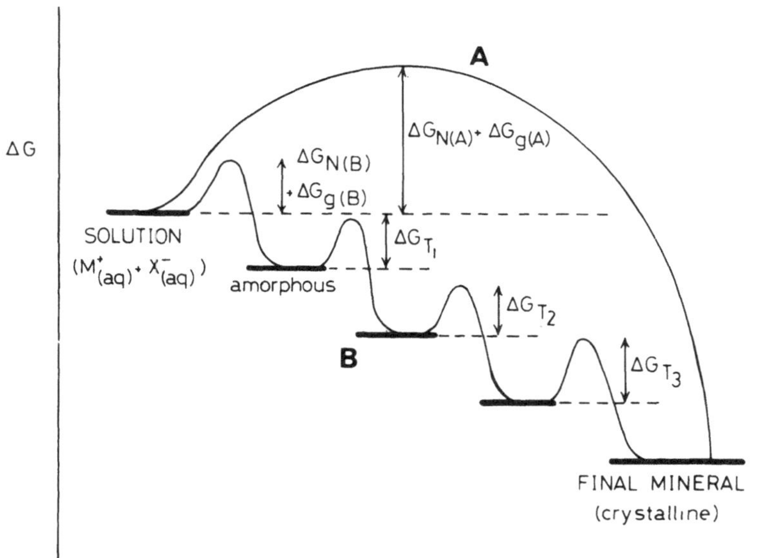

Fig. 3.9 A, B. Schematical representation of the free energy states and the activation energy barriers for mineralization. Pathway **A,** crystallization of a mineral phase from pure and impure solution (with no major structural modifications); pathway **B,** formation of a crystalline mineral from intermediate phases of different crystal structure. ΔG_N, free energy of nucleation; ΔG_g, free energy of growth; ΔG_T, free energy of phase transformation

processes. Figure 3.9 shows schematically two possible mechanisms of mineral formation. Pathway *A* (Fig. 3.9) represents schematically the sum of the activation energies required for the precipitation of a crystalline mineral from pure solution. The energies required to be surpassed are those for nucleation and crystal growth as discussed previously. The formation of crystalline minerals from impure solution involving no intermediates and only minimal crystallographic modification through lattice substitutions can also be represented by pathway *A* provided that the activation energy for crystal growth is subsequently modified. Pathway *B* represents the energy states and activation energies involved in the formation of a crystalline mineral from intermediate phases of different crystal structure. In this case there are many energy barriers of nucleation, growth, and transformation to be overcome. The choice between pathways *A* and *B* and between the possible polymorphs in pathway *B* depends on the relative magnitudes of the corresponding activation energies required for precipitation and phase transformation. Each of these processes can be modulated by biological control of the chemistry of the mineralizing zone.

3.3.1 Chemical Control of Nucleation

The regulation of supersaturation levels at the mineralization site by active ion transport can, in principle, control the rate of nucleation of the biomineral as indicated in Eq. (3) of Sect. 3.1. Also, the choice between pathways *A* and *B* in Fig. 3.9 will be determined to some degree by the structure of the critical nucleus initially formed.

Thus, biomineralization via pathway A (Fig. 3.9) will tend to proceed through critical nuclei of strongly bonded, dehydrated ions essentially arranged as in the bulk mineral whereas mineral formation via an amorphous phase (pathway *B*) will involve nuclei of weakly interacting solvated ions arranged in a diffuse manner. Pathway *B* may be favoured since an empirical observation (the Ostwald-Lussac Law of stages) is that the initial phase formed under conditions of sequential precipitation is the phase with highest solubility. Since amorphous phases are more soluble than crystalline phases at equilibrium it seems possible that the formation of amorphous precursers could be widespread in biomineralization. Similarly, in the sequential precipitation of crystalline polymorphs the more highly hydrated phases will be the first to precipitate, since these crystal forms have the greatest solubility (Table 3.1) due to the higher rate of ion diffusion from the surface because of lower bulk energy terms. Thus for a solution of mineral ions, at any given value of the activity product, $a(M^{n+}) \cdot a(Y^{z-})$ for a solid MY, the amorphous phase of a precipitating mineral will be lower in supersaturation than the crystalline polymorphs. Since the rate of nucleation is related to the supersaturation by Eqs. (3) and (6) (Sect. 3.1), the formation of the amorphous state will only occur if, for any given value of the activity product, the rate of nucleation for crystallization is lower than for the precipitation of the amorphous phase. Thus the following generalised graph can be drawn (Fig. 3.10). This condition can only be true if the higher supersaturation values for the crystalline phase are offset by larger interfacial energies for the critical nuclei involved in the crystallization process. The interfacial energy of a solid-liquid interface (σ_{SL}) can be

Table 3.1. Solubility products for mineral phases occurring in biomineralization[a]

Mineral	Log K_{sp} (25 °C)		Ref.
Calcium carbonate			
Monohydrite	-7.39[b]		37
Vaterite	-7.60[b]		38
Aragonite	-8.22		39
Calcite	-8.42		39
Calcium phosphate			
$CaHPO_4$ (DCPD)	-6.4	-3.2[c]	40
$Ca_4H(PO_4)_3$ (OCP)	-46.9	-6.7[c]	39
$Ca_{10}(PO_4)_6(OH)_2$ (HAP)	-114.0	-7.1[c]	39
$Ca_{10}(PO_4)_6F_2$	-118.0	-7.4[c]	39
Silica			
Amorphous	-2.7[d]		41
Quartz	-3.7		41
Iron oxides			
Amorp. FeOOH	-37.0		40
α-Fe_2O_3	-42.5		40
α-FeOOH	-44.0		40

[a] Values of K_{sp} are limited in application. Activity coefficients and ion-pair formation should be considered where known

[b] Values quoted are calculated activity products

[c] Values of K_{sp} for calcium phosphates in units of mol l^{-1} give a clearer indication of the trends in solubility

[d] Value depends on the type of amorphous silica and electrolytes in solution.

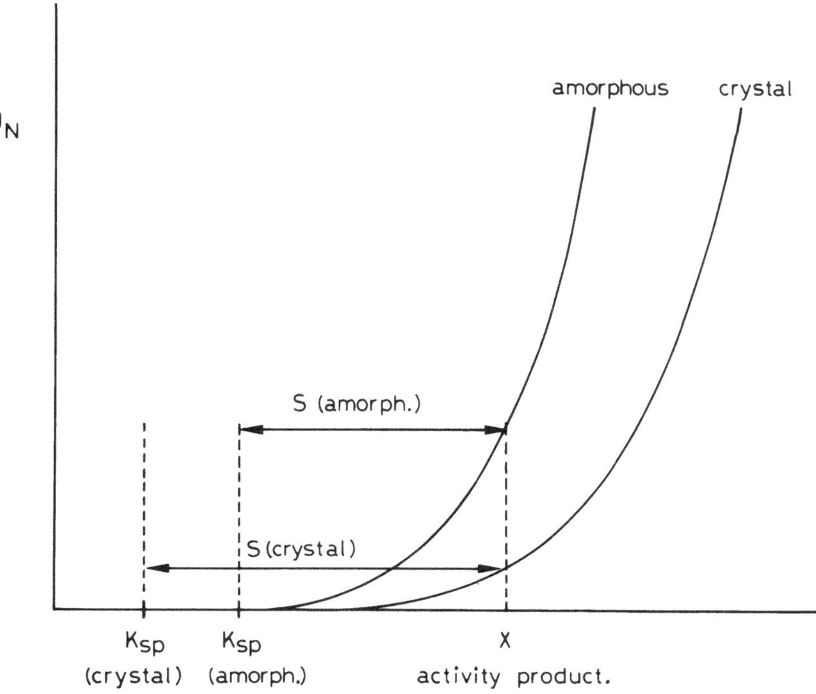

Fig. 3.10. Generalised plot of the rate of nucleation, J_N, against activity product. The solubility product for a possible crystalline phase, K_{sp}(crystal), is lower than for a possible amorphous phase, K_{sp}(amorph.). Thus for any value of the activity product, e.g. at X, the supersaturation for the crystalline phase, S(crystal), is always greater than for the amorphous precipitate, S(amorph.). However, the amorphous phase is kinetically favoured, see text

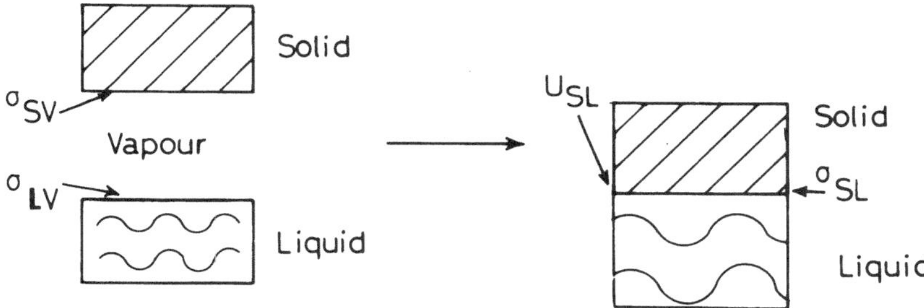

Fig. 3.11. Schematic diagram of the formation of a liquid-solid interface (see text for details)

expressed in terms of the interfacial energies of the solid-vapour interface (σ_{SV}), the liquid-vapour interface (σ_{LV}), and the sum of the solid-liquid interfacial forces per unit area of the interface (U_{SL}) as shown in Fig. 3.11. The solid-liquid interfacial energy, for a reversible system in which the solid is at equilibrium with its surroundings, is then given by

$$\sigma_{SL} = \sigma_{LV} + \sigma_{SV} - U_{SL} \tag{7}$$

σ_{SL} will be larger for a crystalline, rather than a hydrated amorphous phase of a given precipitate since both σ_{SV} and U_{SL} may act to increase this term, with σ_{LV} being constant for the chosen aqueous system. The value of σ_{SV} will be greater for crystalline phases since the lattice energy of this phase will be greater than that for an amorphous phase and there is a direct proportional relationship between the surface energy of a crystal and the value of the lattice energy[12]. σ_{SL} may also be increased for crystalline phases since the sum of the solid-liquid interfacial forces may be lower for well ordered surfaces than for dis-ordered, random ionic arrangements, most probably due to the high energy of dehydration required for the incorporation of solute ions into the crystal surface. A more detailed discussion of this point is given by Williams[102].

Thermodynamically more stable crystalline phases will be favoured as the nuclei grow in size since the bulk lattice energy terms, rather than the surface energy terms, become more important in stabilising the solid phase, particularly (as in the case of amorphous $CaCO_3$) when the amorphous phase is relatively soluble. In cases where the amorphous phase is relatively insoluble, for example, amorphous calcium phosphate, the transformation to crystalline states may be slow at ambient temperature and pressure. In other cases, for example, biogenic silica, the amorphous phase is metastable and not transformed into a crystalline phase under normal conditions due to the high activation energies required to be overcome for this transformation[42].

Alternatively, the volume of the critical nucleus may favour the amorphous state. The number of ions in the critical nucleus will be of the order 10–1000 resulting in a cluster size 5–20 Å. If the stable critical nucleus is smaller than the dimensions of the unit cell of a potential crystalline phase, for example, the large unit cell of hydroxyapatite, then since the nucleus grows without reference to the basic structural unit of the crystalline phase the less-ordered atomic arrangement of the amorphous phase may be favoured.

Note that if precipitation proceeds via pathway *B* of Fig. 3.9, and the final mineral is crystalline, then nucleation will not be a determining factor in the crystallographic structure of the final mineral since the initial phase is amorphous. In such cases mineral structure will be dependent on biological control (chemical and/or organic matrix-mediated) over mineral growth and phase-transformation processes.

3.3.2 Chemical Control of Crystal Growth

In cases where nucleation is unimportant in controlling crystal chemistry the determination of biomineral structure must be governed by biologically controlled physico-chemical processes of mineral growth. Modification of the activation energy barriers for mineral growth and phase-transformation by biological chemical control at the site of mineralization can, in principle, result in the selection of specific crystalline intermediates as shown in Fig. 3.9. Levels of supersaturation at different stages in the mineralization process can be controlled through the regulation of ion concentration gradients across biological membranes. The degree of control that an enclosing membrane can exert has been shown by simplified model systems of intravesicular precipitation in synthetic phosphatidylcholine unilamellar vesicles[43]. These vesicles contained no active transport carriers yet served as an efficient mechanism for controlling intravesicular pH and hence for controlling intravesicular precipitation of inorganic oxides[44, 45]. For example, it has been shown that the intravesicular precipitation of Ag_2O does not occur unless a membrane pH gradient of 4.5 units is surpassed[46]. In these systems, the control of intravesicular concentrations by the slow rate of membrane diffusion of reactant OH^- ions resulted in low intravesicular supersaturation levels which favoured the slow growth of single-domain crystallites in the vesicles. Alternatively, multi-domain and amorphous intravesicular materials were favoured under conditions of rapid membrane diffusion resulting in moderate/high supersaturation levels and nucleation at myraid sites on the inner membrane of the vesicle followed by rapid growth[47].

Chemical control over mineral structure will also be possible through lattice modification by coprecipitating ions and growth regulation by surface active molecules and ions. The growth of pure biogenic crystals seems most unlikely from the necessity of a pure crystallising solution. Most biogenic minerals will be precipitated in the presence of some of the following ions;

Na^+, K^+, Mg^{2+}, Ca^{2+}, Fe^{2+},

OH^-, Cl^-, HCO_3^-, SO_4^{2-}, HPO_4^{2-},

It is therefore most probable that trace concentrations of these ions are incorporated into bulk biomineral phases resulting in modifications in lattice dimensions and physical properties. Some of these ions would be inhibitory to crystal growth. For example, the deposition of biogenic calcite will require the absence of trace amounts of Mg^{2+}. Compartmentalization is a suitable method for attaining these conditions. The investigation of trace ions in biological minerals is limited by the analytical techniques available. The recent developments in proton-induced X-ray emission spectroscopy (PIXE)[48] should allow sensitive detection of elements in localised regions (*ca.* 1 μm^2) within biological minerals.

An extremely important factor in controlling mineral growth and morphology is the presence of accelerators/inhibitors at the mineralizing site. Table 3.2 shows some of the known biological inhibitors of hydroxyapatite (HAP) formation. All these species appear to inhibit the transformation of amorphous calcium phosphate (ACP) to crystalline intermediates. Their inhibitory action can, in the case of phosphometabolites and protein inhibitors, be overcome by the addition of enzymes for which the inhibitors act as natural substrates[53]. Other data shows that some of these inhibitors also act in inhibiting the transformation of crystalline OCP to HAP[58]. In calcification, it has been shown that the formation of aragonite can be stabilised by glutamic and aspartic amino acids through the formation of a protective overgrowth on the aragonite surface via carboxylate groups[59]. Thus the use of inhibitors in determining mineral structure can be highly selective through a process of inhibition, re-initiation, inhibition and so on. A choice between vaterite, aragonite, and calcite, or between the polymorphs of calcium phosphate can then be made, in principle, through the extent of this sequence within the biological system (Fig. 3.12). The modification of this sequence with time and biological location could then result in the development of different crystallographic structures during the stages of growth and thus different minerals may appear in different regions of the mineralized zone. For example, the mature shells of land and freshwater gastropods are composed of aragonite whereas the newly-formed deposits have vaterite structures[60].

Inhibition of mineral transformation by biological molecules is thus potentially a very selective process. Often only a trace amount of the inhibitor will be required to prevent

Table 3.2. Biological inhibitors of hydroxyapatite formation from aqueous solution

Inhibitor	Ref.
Mg^{2+}	49, 50
CO_3^{2-}	49
pyrophosphate ($P_2O_7^{4-}$)	51
polyphosphates; EHDP	51, 52
Cl_2MDP	51, 52
nucleotide polyphosphates;	
adenosine triphosphate	53
guanosine diphosphate	53
glucose 1, 6, diphosphate	53
cartilage proteoglycans	54
dentine phosphoproteins	53
polycarboxylates, $[RCO_2]_n$;	
polyglutamate	55
polyacrylate	55
phospholipids	56
3-phosphocitrate; $[(CH_2CO_2)C(CO_2)OPO_3]^{4-}$	57

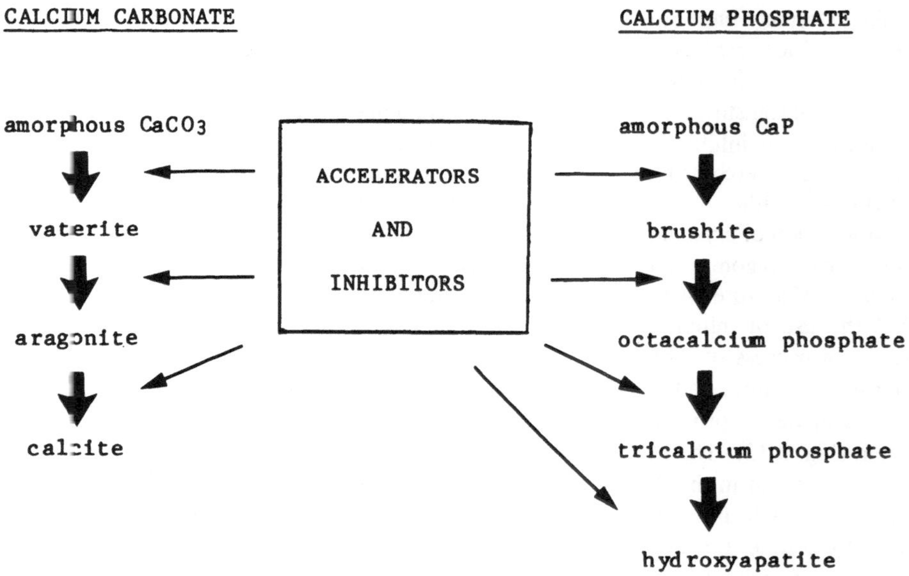

Fig. 3.12. Transformation and regulation of mineral phases under kinetic conditions of biomineralization

crystal growth through blockage of surface sites. Biological design of suitable inhibitors could also result in morphological control. For example, the presence of biopolymers (Fig. 3.13) in supersaturated solutions of sodium chloride inhibited surface nucleation by adsorption onto edge sites allowing dislocation growth to dominate over edge nucleation, and resulted in crystals of well defined cubic habit less susceptible to attrition than crystals in the absence of the inhibitor[61]. In a similar manner, blockage of certain crystal faces could result in crystal growth in preferential directions. Morphological modification of cubic habits resulting in octahedral forms is not unusual. Thus the adsorption of urea on the (111) faces of sodium chloride results in the appearance of higher index faces (Fig. 3.14). Adsorption of such molecules is likely to be greatest for high surface energy planes, particularly if the planes are unicharged (e.g. the (111) faces of NaCl) and have an ionic arrangement which coincides with the configuration of the adsorbing phase. Similarly, habit modification of cubic systems to acicular forms is possible. For example, sodium chloride crystals in the presence of small amounts of polyvinyl alcohol show this habit modification[62].

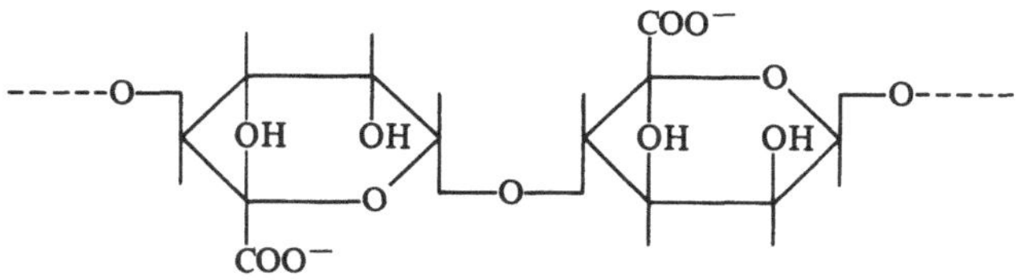

Alginate

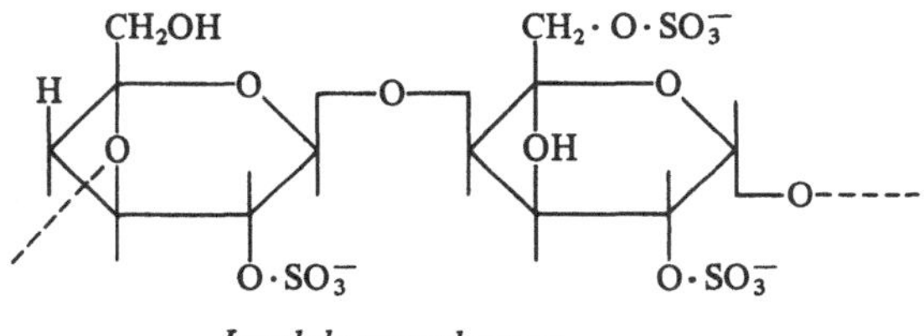

Lambda carragheenan

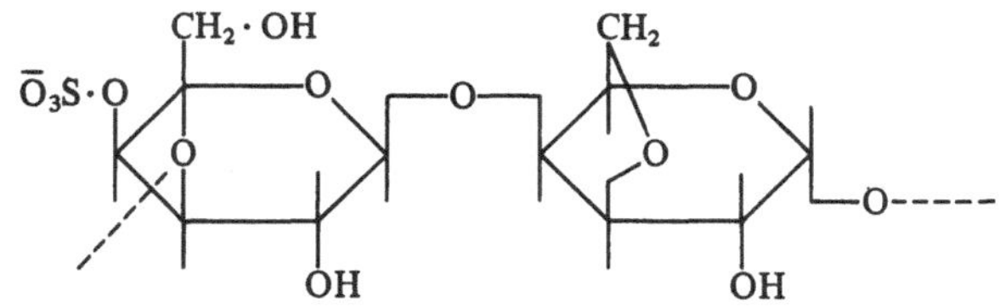

Kappa carragheenan

Fig. 3.13. The structures of sodium alginate and carragheenans used in the inhibition of sodium chloride crystallization from aqueous solution[61]

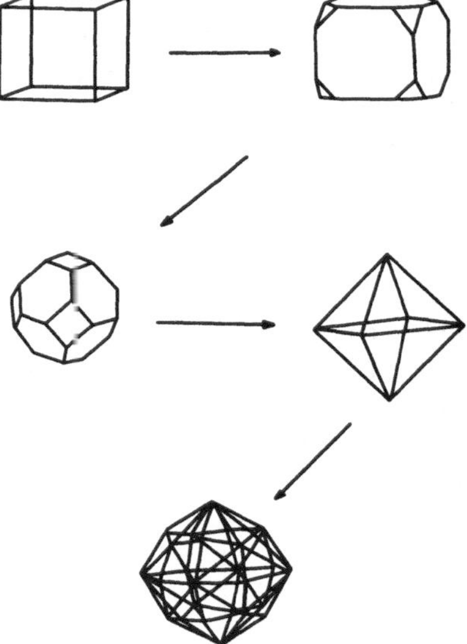

Fig. 3.14. Possible modifications of a cubic system showing first the formation of the octahedral faces and then the higher index faces[12]. The adsorption of urea on the (111) faces of sodium chloride results in this modification

4 Organic Matrix-Mediated Processes of Biomineralization

An organic matrix can be defined as any localised surface comprising organic constituents such as lipids, proteins and carbohydrates which acts as a mediator of mineralization in biological systems. There may be several functions of the organic matrix as summarised in Fig. 4.1. Some functions may be highly specific in the determination of the spatial organisation of mineral particles, crystallographic orientation and mineral structure. Alternatively, the matrix may play a passive role, for example, as an inert substrate for structural support or as a limiting volume for mineral deposition. The matrix will not be strictly inert since a general function will be to act as a surface on which heterogeneous nucleation can occur. The possible reduction in nucleation activation energy will be dependent on the strength of the interaction of the mineral ions in solution and the substrate surface. Chemical bonds between substrate and ions will favour a significant lowering in the nucleation activation energy permitting a moderate rate of nucleation at relative low levels of supersaturation. It is important to realise that this consideration does not necessarily imply that nucleation on substrates of matching atomic lattice dimensions is favoured over nucleation on ill-defined substrates (a condition often implicit in reports of biomineralization), although work on the nucleation of ice on inorganic dust has suggested that lattice matching did have a favourable effect on nucleating ability in these inorganic systems[63]. Hence, this essentially passive function of the organic matrix (with regard to mineral structure and orientation) may be a fundamental role of such surfaces with mineral structure and direction of growth being determined solely by the chemical properties of the mineralizing zone.

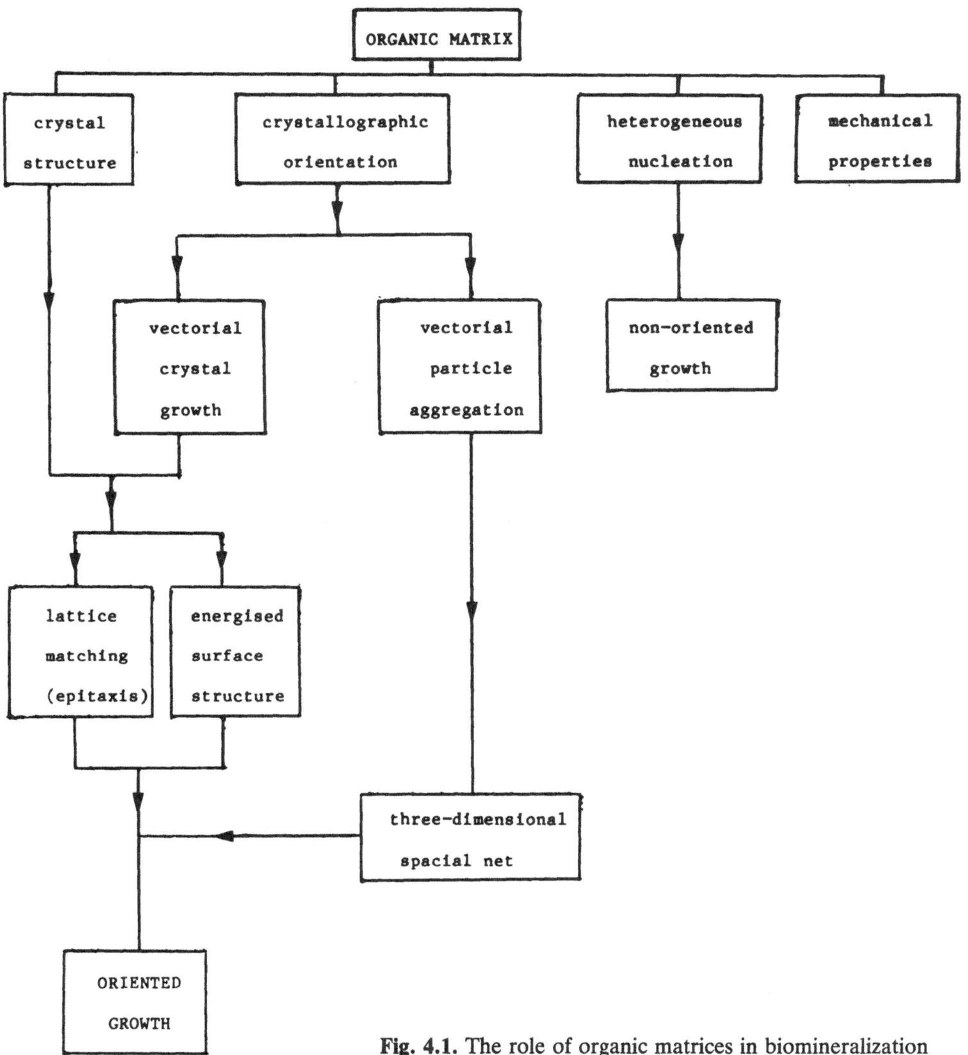

Fig. 4.1. The role of organic matrices in biomineralization

4.1 Spatial Control of Biomineralization

Organic matrices can be utilised in the spatial control of biomineralization in two processes; (i) as a limiting volume for mineralization, (ii) as a surface for controlling the vectorial properties of mineral formation, i.e. crystal size and orientation.

4.1.1 Volume Control of Biomineralization

Volume control over mineralization requires the confinement of the mineralizing zone within a localised volume without preferential alignment of the constituent mineral particles. Control of this kind can be readily achieved by two processes; (i) spatial organic nets, usually extracellular, and (ii) intra- and extracellular compartmentalization through the formation of localised spherical membrane-bound vesicles. In each case the mineral size can be predetermined by the spatial organisation of the biologically-designed compartment. The production of uniform-sized crystallites by this method may be important in determining the mechanical properties of the mineral system.

4.1.1.1 Spatial Organic Nets

Although this paper is primarily concerned with the importance of organic frameworks in the genesis of biominerals it is pertinent to note that such frameworks can be functional in the fully formed mineral. It has long been recognised that organic frameworks consisting of biological macromolecules intertwined in geometric arrays can serve as structural support systems for biological solids[64]. The resulting mosaic biominerals must, even in the simplest support system, resist the forces of tension, compression and bending. The optimization of space frames has been discussed in detail elsewhere[65]. If we consider that, for skeletal structures, the property of least-weight (i.e. the use of minimal material in order to perform a given function) is the fundamental requirement in optimizing the structure, the theoretical results show that the optimum structures reduce to orthogonal nets (Fig. 4.2)[65]. The simplest case is the rectangular net (Fig. 4.2 a) but the results show that the equiangular spiral (Fig. 4.2 b) and the circular fan (Fig. 4.2 c) come very close to absolute minimum weight except for restrictions at the origin of the spiral and at the closure of the fan where strains may be incompatible. A further conclusion of the optimization theory is that the optimum framework of minimum weight is also the stiffest of all possible frameworks whose members sustain tension and compression stresses in that region of space. It has been suggested[64] that biological frameworks are often not designed at this optimum level implying that they may be intentionally less stiff, i.e. they are designed to accommodate rather than resist deformations.

The rigid structure of spatial nets can be formed by strands of connecting tissue such as long multiple helices of protein. The most basic structural fibre of the animal phyla is collagen which has been investigated in great detail[66, 67]. The proteins in collagen are composed mainly of simple amino-acids, glycine, proline, and alanine. Occasionally this sequence is broken by isolated runs of more polar amino-acids or by large hydrophobic

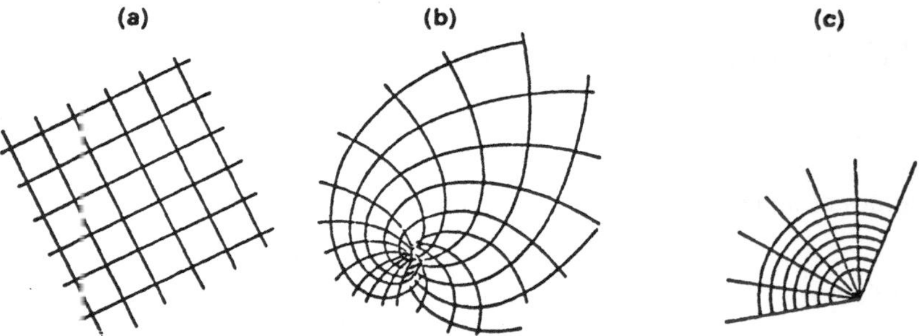

Fig. 4.2 a–c. Possible arrangements for spatial nets[65], **a** rectangular net; **b** equiangular spiral; **c** circular fan

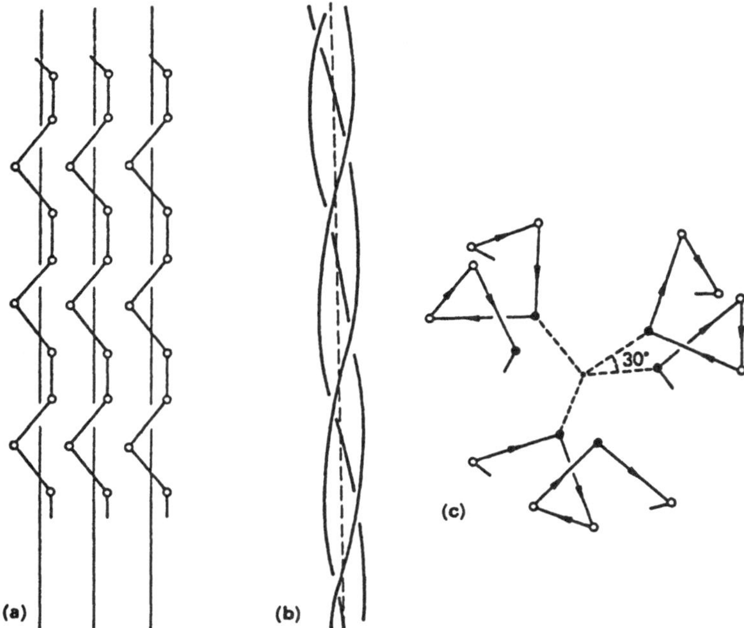

Fig. 4.3 a–c. The coiled-coil structure of collagen. **a** three polypeptide chains coiling around three separate (minor) axes; **b** the three minor axes coiled around a central axis (*dashed line*); **c** end-on view of the coiled-coil. Each circle represents the α-carbon of an amino acid. *Solid circles* are glycine. *Solid line* in **c** is the polypeptide chain[68)]

residues. The structure of collagen is based on a helical arrangement of three, non-coaxial, helical polypeptides, and stabilised by inter-chain hydrogen bonding (Fig. 4.3)[68)]. This arrangement is considered to be the basic subunit for the assembly of the collagen fibre which can take place by a variety of permutations depending on the chemical composition of the precipitating medium[69)]. One arrangement, the "revised quarter-stagger" model[70)] has a structure with 64 nm periodicity and is typical of vertebrate and numerous invertebrate collagens (Fig. 4.4). Each molecule is divided into five zones, the first four zones being of equal length (64 nm) and the fifth zone only 25 nm. This arrangement requires a 30–40 nm space between the ends of the molecules. The specific arrangement of these spaces ("hole zones") may play an important role in the mineralization of collagenous tissues such as bone, as described below (Sect. 4.1.2.1).

An example of volume control by the secretion of an extracellular three-dimensional organic net followed by the precipitation of the mineral phase within this framework has been observed in magnetite (Fe_3O_4) mineralization in the radular teeth of chitons[71)]. Mineral deposition begins at the edges of a three-dimensional fibrous net (Fig. 4.5 a) and gradually proceeds to fill in the interspaces within the organic framework (Fig. 4.5 b). The magnetite crystallites are aggregated within the interspaces such that there is random crystallographic orientation. The organic framework then acts as a matrix for volume control without affecting the crystallographic or crystallochemical properties of the mineral phase.

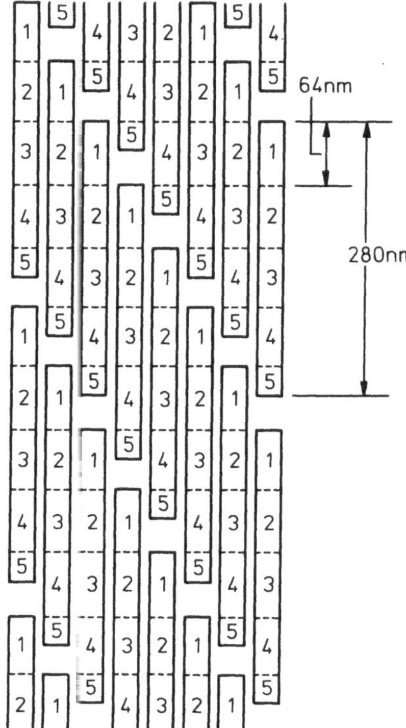

Fig. 4.4. The "revised quarter-stagger" model for collagen (in planar form)[70]

4.1.1.2 Membrane-Bound Vesicles

Intra- and extracellular vesicles have been observed to be compartments for mineralization in many biological systems. A major use of vesicles in the volume control of biological precipitation is in pathologically-induced processes such as detoxification. Thus high concentrations of toxic heavy metal ions such as Cd^{2+}, Zn^{2+} and Pb^{2+} are rendered inert by precipitation in membrane-bound vesicles of the kidney cells of marine molluscs[72, 73] followed by excretion in the urine[74]. Similarly, pollutant metals are concentrated as trace elements in Ca/Mg pyrophosphate membrane-bounded granules in the hepatopancreas tissue of the common garden snail[75]. Volume control is also important in magnetite (Fe_3O_4) mineralization in magnetotactic bacteria. Precipitation occurs within membrane-bounded vesicles resulting in crystals of optimum (single-domain) dimensions for remanent magnetization[8] (Fig. 4.6).

More sophisticated volume control can be achieved by the biological design of vesicle geometry. More energetically-demanding geometries are possible, for example, in silicification. In the choanoflagellate, *Stephanoeca diplocostata* Ellis curved silica rods of ca. 3 μm length are constructed within preformed curved elongated vesicles[5] (Fig. 4.7). Similarly, in diatoms, the geometric patterning of the shell (frustule) is the result of localised amorphous silica deposition within vesicles of predetermined shape and size[9]. A similar mechanism of shell patterning may occur in the formation of the calcified plates of coccoliths[6].

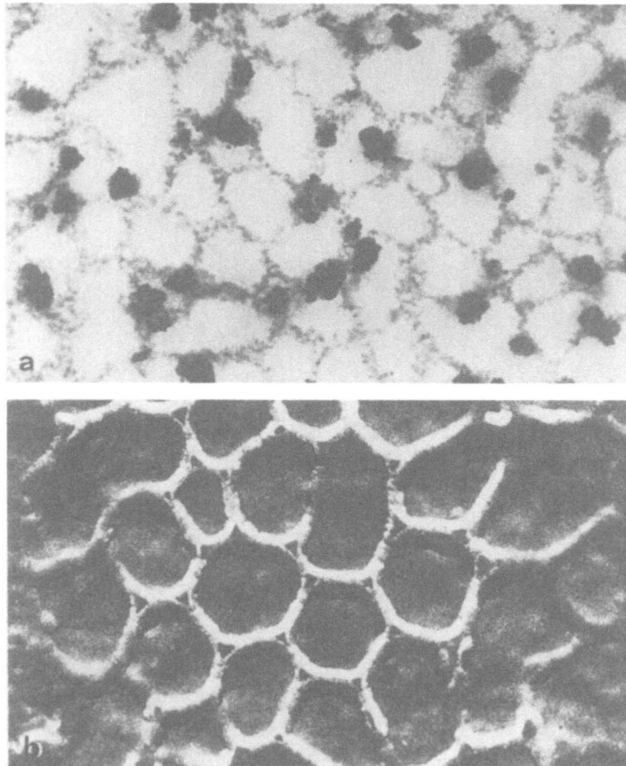

Fig. 4.5 a, b. Fe_3O_4 mineralization in the radular teeth of chitons[71]; **a** initial stages of mineralization showing Fe_3O_4 crystallites deposited on a fibrous organic framework (× 80,000); **b** late stage in mineralization showing the polygonal organic framework enclosed in solid Fe_3O_4 (× 80,000)

4.1.2 Control of Vectorial Properties of Biomineralization

There are two possible mechanisms of controlling both the orientation and size of mineral particles; (i) vectorial particle aggregation, and (ii) vectorial crystal growth.

4.1.2.1 Vectorial Particle Aggregation

Bulk orientation of a biomineral can arise from the ordered alignment of mineral sub-units. Vectorial particle aggregation implies that these inorganic sub-units are of controlled size and are then aggregated in preferential directions. Crystallite size could be determined before aggregation by precipitation within limited biological volumes such as in the compartments of vesicles. Transportation of the vesicles to an extracellular organic matrix followed by spatial ordering of the released mineral particles on the organic surface could then result in oriented growth of the bulk mineral phase. In general, vectorial particle aggregation is a process of ordered flocculation or coagulation where

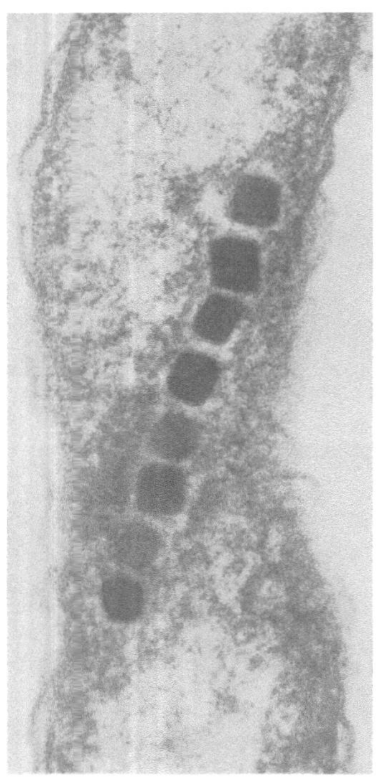

Fig. 4.6. Thin section out from a magnetotactic bacterium showing magnetite (Fe_3O_4) crystals enclosed within intracellular vesicles[8] (× 152,000)

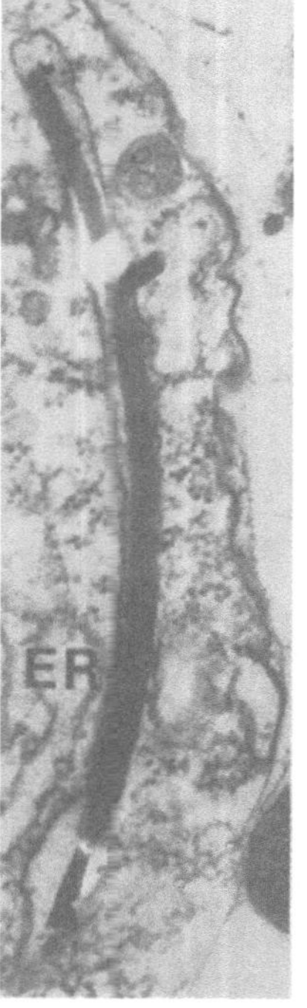

Fig. 4.7. Sectioned cell of *Stephanoeca diplocostata* Ellis showing a mature intracellular silica costal strip enclosed within a preformed elongated vesicle adjacent to the endoplasmic recticulum (ER)[5] (× 34,000)

the inorganic crystallites are linked together by bridges of the surfactant. The interaction between the crystallites and flocculating agent will arise by adsorption of the flocculating agent through ionic and hydrogen bonds at the crystallite surfaces. Spatial alignment of the crystallites will occur through chemical interactions between the charged crystallite surfaces and oppositely charged side chains of the polymeric framework. The resulting mineral is then oriented with respect to an organic matrix without the necessity of lattice matching (see Sect. 4.2) between mineral and substrate. Extension of this structure in three dimensions will require the continual synthesis of the organic matrix such that the bulk mineral will be a three-dimensional mosaic of crystallites interspaced with organic material. The mineral may appear macroscopically as a single crystal if there is preferential orientation of the crystallites resulting in an iso-oriented mosaic structure.

The process of vectorial crystal aggregation plays no role in the determination of the structure or morphology of the crystallites comprising the bulk mineral since it orientates preformed particles of determined size in space. Crystallite structure and morphology must thus be determined by localised chemical control over the formation of the discrete mineral particles prior to spatial alignment.

It is conceivable that a process of vectorial particle aggregation could be utilised in the macroscopic ordering of amorphous particles such as in the biological mineralization of silica (SiO_2) since ordered aggregates of this kind are known in inorganic systems such as in "precious" opal. These materials are composed of a regular packing of silica particles of uniform size resulting in a macroscopic lattice structure. The lattice spacing of these structures is comparable to the wavelength of visible radiation and results in highly coloured materials when observed from certain directions of the reflected light due to diffraction interference.

A probable example of vectorial crystal aggregation can be found in the biomineralization of calcium phosphate. With the exception of enamel apatite, the vast majority of the microscopic (0.1 μm) crystals of vertebrate apatite are observed to be embedded in an extracellular, largely collagenous, matrix. These crystallites are arranged at a late

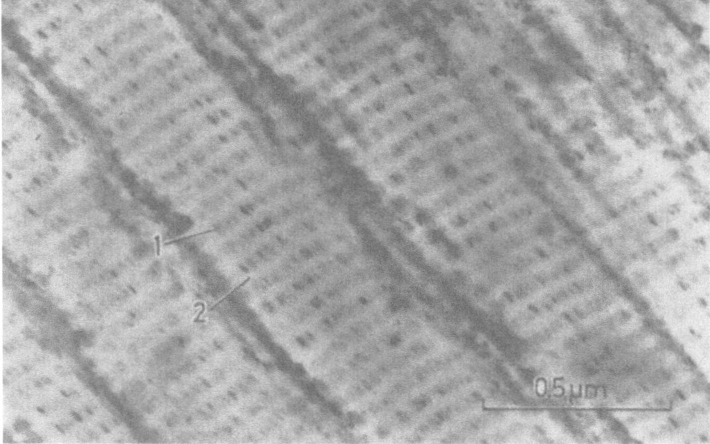

Fig. 4.8. Electron micrograph of mineralized turkey leg tendon showing possible vectorial ordering of apatite crystallites on a collagenous matrix[76] (× 50,000); *arrow* (*1*), arrangement of crystallites in lines in the "hole" zone; *arrow* (*2*), "overlapping" zone free from crystallites

stage in the calcification process in densely packed, highly ordered arrays[76] (Fig. 4.8). Initiation of the solid phase takes place apparently in membrane-bound spherules (matrix-vesicles)[77–79] which are extruded from the parent cell plasma membranes to the extracellular space in which there is a preformed organic matrix. The first solid phase to be precipitated within the vesicles is amorphous calcium phosphate[80, 81], which slowly transforms to apatite crystals inside and on the surface of the vesicles. Islands of coalesced crystallites are next observed in the extracellular-matrix[82] and the total matrix space is eventually filled with the crystallites. The crystallites are spatially ordered at the matrix surface with the apatite *c* axis aligned with the long fibre axis of the matrix[76]. Ordered aggregation of these crystallites on the collagenous fibres is then a probable mechanism of calcification since the aggregation of the vesicular precipitates (whose crystallochemistry, morphology, and size will be determined by the localized chemical environment of the vesicles), could be controlled by the specific spatial organisation of the collagen fibres (Sect. 4.1.1.1). However, it must be noted that another possible mechanism is that of vectorial crystal growth (Sect. 4.1.2.2) since it is not totally clear whether the mineral front is laid down by the direct ordering of the intravesicular deposits or by the dissolution of the vesicular crystallites followed by ion translocation and reprecipitation at the collagen surface.

4.1.2.2 Vectorial Crystal Growth of Biominerals

The growth of minerals on specific crystal faces such that the crystals are of controlled size and oriented in preferential directions can occur through the deposition of ions on an active polymer surface. Growth on specific crystal faces may be favoured by many different properties of the organic interface. Crystallographic orientation can result from an atomic matching of substrate and overgrowth or without lattice matching at the interface. In the former case, the atomic relationships between the substrate and the crystal faces of the overgrowth energetically favour preferential growth on faces at which there is the closest possible atomic matching. In the latter case, the crystallographic orientation of the overgrowth is determined by the surface geometry of the spatial frame. Thus the molecular configuration of polypeptide side-chains can result in surface irregularities at the interface which energetically favour nucleation and growth along particular crystal faces of the overgrowth. The size of the crystallographically oriented mineral sub-units can be determined in both these processes by the temporal and spatial regulation of the synthesis of the organic matrix. Thus layers of the oriented mineral can be constructed sandwiched between lamellae of organic material.

A probable example of biological control over the crystallographic orientation and size of mineral growth is in the formation of calcitic otoconia in mammals. Investigation of the early stages of otoconia formation[83] has shown that mineralization occurs by precipitation of ions from the surrounding endolymph fluid onto the surface of an extracellular organic matrix. The crystals have been deduced to be single calcite crystals on the basis of X-ray diffraction data[84]. However, as Fig. 4.9 shows, these crystals are themselves composed of a multitude of small discrete crystallites aligned in preferential directions[85]. Recent investigations using ultra-high resolution electron microscopy has shown that these biominerals fracture along different faces, not necessarily those found for their inorganic counterparts, revealing microscopic (100 nm) single-crystal sub-

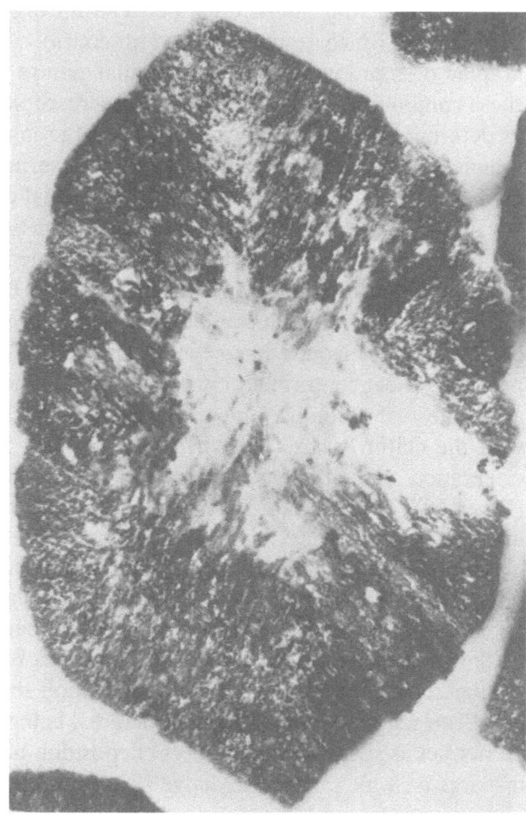

Fig. 4.9. Sectioned mouse otoconium showing an iso-oriented mosaic of calcite sub-units within an organic matrix[85] (× 40,000)

units[86]. It has been postulated[85, 86] that the nucleation and oriented growth of these subunits is controlled by the molecular/atomic properties of the organic surface. Thus, even though the bulk mineral has the morphology of a single crystal the internal structure is very different being composed of an iso-oriented mosaic of crystalline subunits.

4.2 Structural Control of Biomineralization by Organic Matrices

A major concept in biomineralization is the utilisation of an organic matrix as a means of attaining specificity in mineral crystallochemistry. Structural control of the mineral phase can be achieved by the biological design of two-dimensional atomic nets which act as lattice templates for mineralization on organic substrates (epitaxis) or by the interaction of depositing ions on a molecular-specific organic surface without the requirement of a lattice match between the substrate and overgrowth. Thus, in organic matrix-induced epitaxis the atomic match at the interface determines the crystallographic structure of the forming mineral and any modification of the two-dimensional organic net implies a corresponding change in mineral crystallochemistry. Determination of the depositing polymorphs can then be achieved through control of the crystal space group, e.g. between aragonite (orthorhombic) and calcite (hexagonal).

In general, the interactions between an organic surface and depositing ions at an atomic level which result in structural control will also result in control over crystallographic orientation. Thus the molecular nature of the organic interface can be, under these conditions, as important as the macroscopic spatial arrangement of the framework in determining the spatial orientation of the mineral phase. The function, then, of an atomic net is to provide a two-dimensional surface which favours nucleation of specific crystal forms and crystal growth in preferential directions.

Whether epitaxial processes are dominant in biomineralization is open to debate although crystallochemical control by such mechanisms has been considered to be a major mechanism in some mineralizing systems. For example, a correspondence between the crystal lattice of the insoluble organic matrix isolated from mollusc shells and the unit cell lattice of aragonite has been suggested[87)]. The matrix consists of a protein of anti-parallel β-sheet conformation with a polysaccharide phase of chitin. The 5 Å *a* axis of the aragonite orthorhombic unit cell was then matched in orientation and lattice spacings with the chitin *b* axis (protein *a* axis) and the *b* and *c* axes of aragonite matched in orientation and lattice spacings with the *b* and *c* axes of the protein respectively (Fig. 4.10). The role of β-sheet structures in epitaxial control over calcification has been discussed elsewhere[42, 88)]. These structures have polypeptide chains high (*ca.* 85%) in glycine, alanine, and serine in the form of pleated sheets linked by inter-chain hydrogen bonds (Fig. 4.11)[89)]. The structures may be parallel or anti-parallel pleated sheets, depending on the disposition of the side chains between the pleated sheets. The extention of the amino-acid side chains into the spaces between the pleated sheets results in a regular repeat pattern of potential nucleation sites for mineralization. However, it has also been suggested that β-sheet structures, being relatively inert to chemical influences, do not act as substrates for mineral deposition but as substrates for over-laying organic matrices on which there are active sites for mineralization[60)].

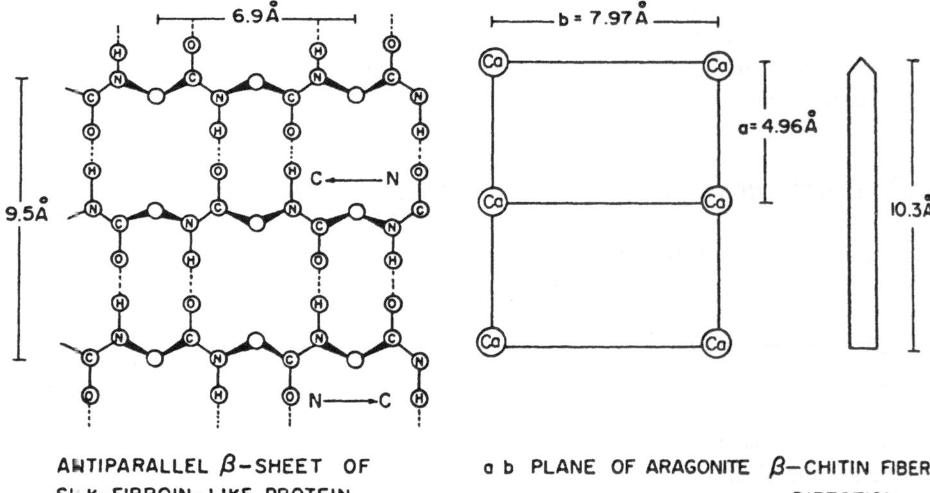

Fig. 4.10. Schematic representation of the structural relationships between protein sheets, aragonite crystals, and chitin fibres in the nacreous layer of *N. repertus*[87)]

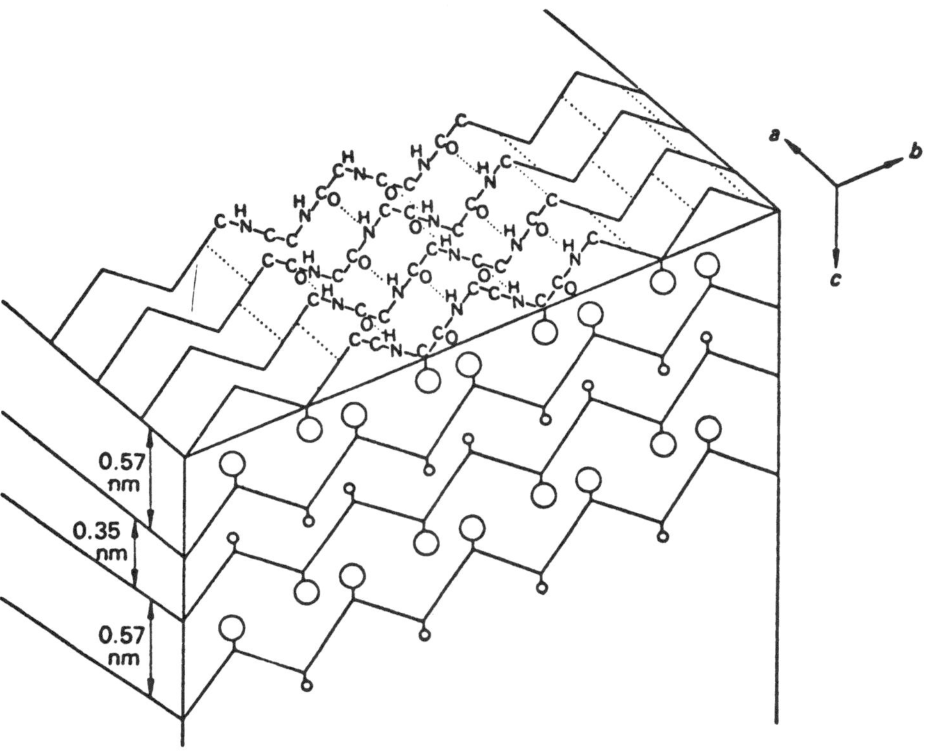

Fig. 4.11. Pleated *β*-sheet model of *Bombyx mori* silk[89)]

The possibility of structural control arising from non-epitaxial processes has not generally been considered even though there are many examples of oriented overgrowth in inorganic systems in which lattice matching appears to be unimportant. Unfortunately, one of the major problems in biomineralization lies in the identification and characterisation of the organic phase *in vivo*. The isolation of proteins from biomineral deposits followed by sequence-analysis is, in principle, a method which should provide information about the atomic arrangement of organic matrices. Unfortunately this approach may be limited since the matrices are likely to be composed of many different proteins which are intricately related *in vivo* and which bear little resemblance to their interactions studied *in vitro*. However, as discussed in Sect. 4.3, there are several *a priori* grounds for considering non-epitaxial, rather than epitaxial deposition, as the major mechanism for the formation of mosaic minerals in biology.

4.2.1 Structural Control by Epitaxis

Epitaxis has been studied in great detail for inorganic crystals formed on insoluble substrates (Table 4.1). Epitaxial deposits can be divided into two groups, those in which orientation is controlled by initial nucleation and those in which orientation is controlled

Table 4.1. Epitaxial deposition of inorganic crystals on insoluble substrates[90)]

Substrate	Deposit	Lattice Misfit %	Orientation Substrate	Orientation Deposit
PbS	NaI	8	(001)	(001)
	KCl	5		
	NaBr	−1		
	NaCl	−6		
	AgBr	−4		
	AgCl	−7		
$CaCO_3$	RbBr	7	(100)	(100)
	RbCl	3		
	KBr	3		
	NaI	1		
	KCl	−2		
	NaBr	−7		
CaF_2	NaBr	8	(111)	(111)
	NaCl	3		
	LiBr	0		
	LiCl	−6		
NaCl	NaBr	6	(001)	(001)
	NaCN	6		
	AgBr	3		
	AgCN	3		
	AgCl	1		
	KF	−5		

by subsequent growth processes. In both cases, the substrate exerts crystallochemical control over the developing solid phase.

In nucleation-controlled epitaxis, the overgrowth orientation and structure will be determined by the change in free energy of the substrate/overgrowth interface as a function of orientation of the developing crystal faces. Thus epitaxis will occur when the activation energy for nucleation shows a distinct minimum for a specific crystal face caused by the atomistic structure of the substrate surface. The interfacial energy will increase with the misorientation angle and with decreasing supersaturation levels (Fig. 4 12). If heterogeneous nucleation is favoured by epitaxis (it may not be) then the difference between the nucleation probability for epitaxis and non-oriented growth becomes smaller with low supersaturation levels and large lattice mismatch. Therefore, even for organic matrices of specific substrate-mineral lattice match a high degree of supersaturation may still be required in order to favour epitaxis over non-oriented growth.

In growth controlled epitaxis the mineral orientation and structure will be determined through the growth processes of an initially formed lattice-matched monolayer. Thickening of this monolayer with mineral growth will generate dislocations. If these dislocations are parallel to the substrate then the initially oriented layer will be preserved in subsequent layers provided that the dislocated layer is at least a few monolayers thick for these dislocations to be smoothed out. If dislocations do not occur in the early stages of layer

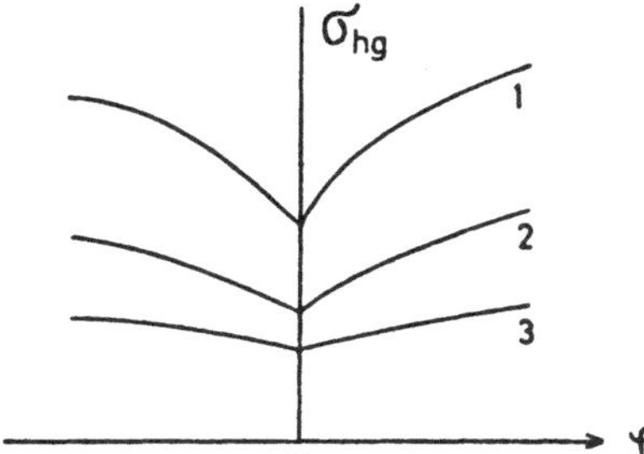

Fig. 4.12. Generalized plot of interfacial energy verses misorientation angle φ at; (*1*), low supersaturation; (*2*) and (*3*), high supersaturation[91]

formation the large strain energy readily tolerated by the interfacial layers will not be possible in the thicker (less readily compressible) films which will result in a strain transition process as the mineral grows on the substrate. The consequence of this growth process is that the thickening of films may be accompanied by a change in the initial orientation on release of the interfacial strain, for example, through slip processes occurring simultaneously in planes inclined to each other[92]. Thus, Ag and Ni films grown on NaCl crystals have a base consisting of small crystallites having (221) faces in contact with the substrate[90]. However, the bulk of the films grown are twinned on the (111) face with an orientation parallel to the NaCl crystals. The corresponding misfit is 9% for Ag, and 7% for Ni in the interfacial region but the bulk misfit is – 27% for Ag and – 38% for Ni, values outside the tolerance limits usually described for epitaxis. Table 4.2 shows several examples of this mode of epitaxial deposition.

4.2.2 Structural Control by Non-Epitaxial Processes

The study of epitaxial deposition in inorganic systems suggests that this process is only one possibility of structural control in biomineralization since it appears that a high degree of lattice matching is not always required for oriented overgrowth in inorganic systems (Table 4.3). In such cases other mechanisms of oriented growth must be appli-

Table 4.2. Epitaxial deposition of inorganic crystals on insoluble substrates in which the final orientation is established through intermediate layers of different orientation[90]

Substrate	Deposit	Lattice Misfit %	Orientation Substrate	Orientation Deposit
NaCl	Ag	-27	(001)	(001)
	Ni	-38		
	Cu	-36		
KCl	Ag	30	(100)	(001)
	Pd	24		
$CaCO_3$	Au	-5	(001)	(111)
	Ag	-5		
	Pd	-10		

Table 4.3. Oriented overgrowth of inorganic crystals on insoluble substrates with large lattice misfits[92)]

Substrate	Deposit	Lattice Misfit %	Orientation Substrate	Orientation Deposit
NaCl	NaI	15	(001)	(001)
	KCN	16		
	NH_4Cl	16		
	KBr	17		
	NH_4Br	23		
	KI	25		
	NH_4I	29		
	RbI	30		
PbS	RbBr	15	(001)	(001)
	KI	18		
CaF_2	KBr	21	(111)	(111)

cable. A high degree of mismatch can be tolerated if the surface layers of the deposit can be distorted to fit the substrate with a one-to-one correspondence. This distortion can be accomodated by a suitable number of dislocations in the interfacial layer which take up the disregister between the bulk crystal and the stretched surface. Another possibility is that the deposited crystallite is not distorted but that there is a bulk disregister between deposit and substrate rather than at the solid-solid interface. Oriented growth can still occur since the energy for growth is likely to be a minimum in some preferred direction even though no direction corresponds to a matching register. For very large degrees of lattice mismatch there will be a high density of dislocations at the interface such that the depositing solid phase will have lattice spacings approximating to the natural unit cell spacings and not that of the substrate lattice. In these cases, the initial monolayer formed at the interface will be mobile on the substrate surface, free to rotate and glide. Thus under these conditions the monolayer can not be an embryo for epitaxial deposition and there is subsequently no substrate control over crystallochemistry by lattice matching.

It has, therefore, been increasingly recognised that oriented overgrowth can arise from a multitude of factors which are related to the thermodynamics and kinetics of the nucleation and growth processes; only one of these factors being the matching of atomic positions at the interface. Thus it has been observed that different orientations of a given crystal overgrowth can arise on the same substrate under fixed experimental conditions[93)]. The structure of the surface of the substrate is then of upmost importance. Thus, although lattice matching may aid oriented overgrowth, other factors such as active sites on the surface can also favour growth on preferential crystal faces. Active sites can result from surface irregularities such as steps, kinks, and dislocations and from the adsorption of foreign ions on the surface. The interaction of ions at the surface resulting in oriented overgrowth can be considered as a system of competition between different modes of deposition with different crystal planes energetically favourable to some surface configuration of the substrate[94)].

The presence of surface active sites will favour strong bonding in these regions. The study of deposition on inorganic surfaces of thin metal films has shown that nucleation is a dynamic phenomenon. The three-dimensional nuclei (20–100 Å) are highly mobile across the surface[95]. Such nuclei are free to rotate and aggregate resulting in oriented layers according to the surface geometry. Thus the orientation of the final mineral may not be the same as that of the initial embryos.

4.3 The Function of Organic Matrices in Oriented Mineral Growth

Oriented overgrowth can be defined as an observational term describing the general phenomena of oriented crystal growth on an insoluble substrate surface without any reference to its genesis. Oriented overgrowth in biomineralization may result from the regular alignment of ions deposited from solution at an organic interface or from the ordered alignment of preformed crystallites in a three-dimensional organic matrix. Oriented overgrowth from the contact solution can result from a matching of substrate and deposit lattices (epitaxis) or from an energy relationship between the surface geometry of the substrate and the depositing nucleus. Figure 4.13 illustrates these three possible mechanisms of oriented overgrowth on organic substrates.

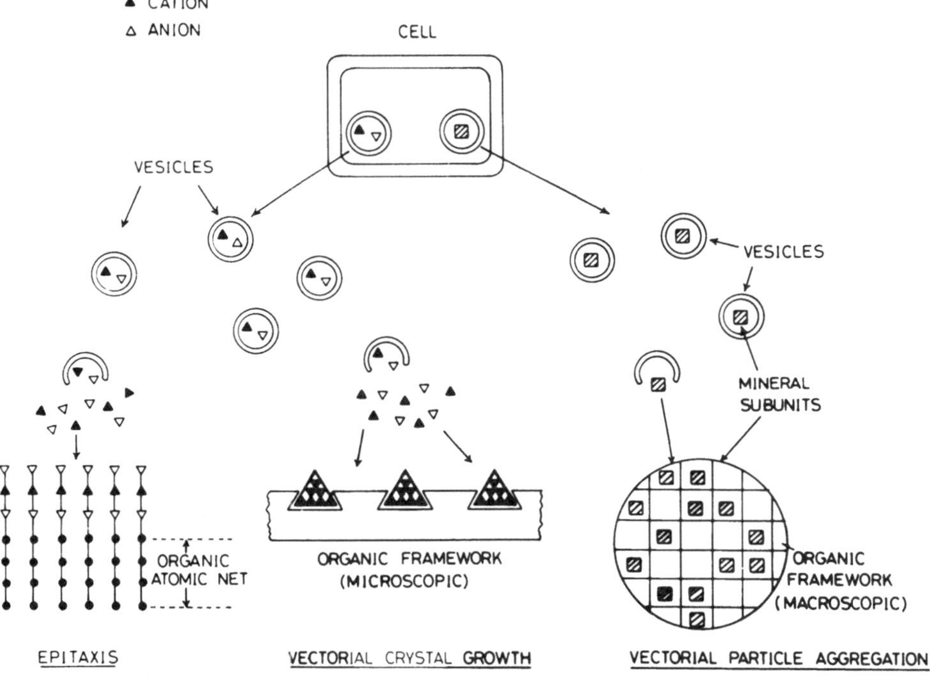

Fig. 4.13. Generalised scheme showing the three possible mechanisms of oriented overgrowth in biomineralization. Oriented growth can arise from ordered aggregation of preformed mineral particles or from the regular alignment of precipitating ions at an organic surface. In the latter case two processes are possible; (a) epitaxis, in which there is a lattice match between the organic atomic net and the lattice spacings of the crystal faces, (b) non-epitaxial processes, in which the surface structure of the substrate determines the orientation of the developing mineral

We must now consider the important question of the probability of these mechanisms taking place in biomineralization. It is often implicitly assumed that the main function of the matrix is to act as a template for mineral development through the control of crystal orientation and structure, but as we have discussed previously this is not necessarily correct since the matrix may serve simply as an energetically favourable site for heterogeneous nucleation, or as a means of volume constraint, or as a framework for structural support. Thus the bulk mineral is often observed to be composed of crystallites interlaminated and wrapped with organic sheets and it is often not at all clear what function the matrices serve.

A further problem is that the observation of bulk oriented overgrowth may not necessarily imply that this relationship originates from an organic matrix-induced process since there is also the possibility of thermodynamically unstable initially non-oriented mineral phases rearranging *in situ* to oriented mineral overgrowths. Thus, the formation of an oriented mineral phase in the presence of an organic framework could arise from the phase transformation of a non-oriented phase by surface dissolution followed by epitaxial deposition of the translocated mineral ions at the surface of the inorganic precursor. The final mineral phase may thus be oriented with respect to the organic substrate without any organic matrix-mediated processes being involved. In such cases the organic substrate may still be advantageous since it can favour the nucleation of the precursor phase and act as a spacial framework for volume control over biomineralization.

An obvious case where organic matrices can not possibly act as templates for oriented growth is in the formation of amorphous biominerals. Since the number of amorphous precursers observed in biomineralization has increased markedly in recent years (Table 1.1) (the difficulty in isolating these phases as precursers may be due to their rapid transformation to crystalline phases *in vivo*) there appears to be some evidence that the control of mineral orientation and structure by atomic relationships between matrix and crystal is not a major role of organic frameworks.

Of the three mechanisms of oriented growth discussed above, vectorial aggregation of mineral particles is likely to be the most readily recognised in practice since the presence of mineral particles may possibly be observed in regions away from the bulk mineralizing site when thin sections of the mineralizing tissue are studied by electron microscopy. However, the discernment between epitaxial and non-epitaxial mechanisms of oriented growth is more difficult since the investigation of the atomic structure of the interface is virtually impossible. Comparison between the bulk mineral orientation and the molecular structure of the isolated organic phase will undoubtably aid our understanding but there is the problem that the orientation of the bulk mineral may not be the same as that of the initial embryo. A possible method for studying the atomic relationship between the organic matrix and mineral overgrowth is by ultra-high resolution electron microscopy provided that samples thin enough for lattice imaging can be obtained.

There are, however, some *a priori* arguments which favour non-epitaxial processes as the major mechanism of oriented overgrowth of biominerals from ion precipitation at organic surfaces. Firstly, epitaxis requires a rigid two-dimensional atomic net whereas non-epitaxial deposition can be achieved from a more flexible surface on which there are specific molecular configurations of polypeptide side chains. For example, it has been suggested that the deposition of apatite within the 640 Å ("hole") zones between collagen molecules takes place by an epitaxial process[96] but n.m.r. data has showed that both the peptide backbone and amino-acid side chain carbons of the collagen fibres display considerable molecular motion[97] indicating that the matrix may not be static enough on a molecular level for epitaxis to occur.

Secondly, many biominerals are observed to be formed within membrane-bound vesicles, often of extremely small size (≤ 100 nm). Epitaxis can occur within the vesicles if nucleation takes place at regularly spaced sites in the membrane surface. These sites could be charged lipid headgroups or active groups of an organic matrix located at the membrane surface to maintain the molecular rigidity required for lattice matching. However, in spherical vesicles the curved membrane surface will limit the extent of epitaxis since crystal growth from different points on a curved surface will result in incoherent development of the mineral nuclei (Fig. 4.14). Thus the formation of an intravesicular single-crystal would be favoured by nucleation at one site on the inner membrane surface. These observations are verified from recent studies of intravesicular precipitation in synthetic, 300 Å diameter, phosphatidycholine vesicles[44, 45, 49]. Metal ions have been encapsulated in these spherical vesicles and the corresponding anions required for solid formation transferred into the vesicle compartment from the external medium either by passive diffusion or by the use of ionophores located in the vesicle membrane. The results showed that single-domain crystallites, polycrystallites, and amorphous materials could be formed within the vesicles. Single-domain crystallites (e.g. Ag_2O) were favoured for metal ions which have low binding constants at the negatively charged phosphate headgroups. In these cases single crystals can be formed by nucleation at one site on the membrane surface followed by slow crystal growth (low supersaturation levels)[45]. Alternatively, metal ions which have high binding constants to the phosphate headgroups favoured the formation of polycrystalline and amorphous intravesicular deposits located around the membrane surface. For example, Co(II) ions formed diffuse amorphous $Co(OH)_2$ intravesicular deposits[98].

Biology can overcome this problem of matrix/vesicle geometry by maintaining the matrix in free solution within the vesicle or by constructing elongated vesicles. In both cases epitaxis can occur since the substrate is presented as a two-dimensional plane rather than as a curved surface. It seems however, unlikely that the intravesicular matrix could be maintained in a static state in free solution within such a micro-volume of space.

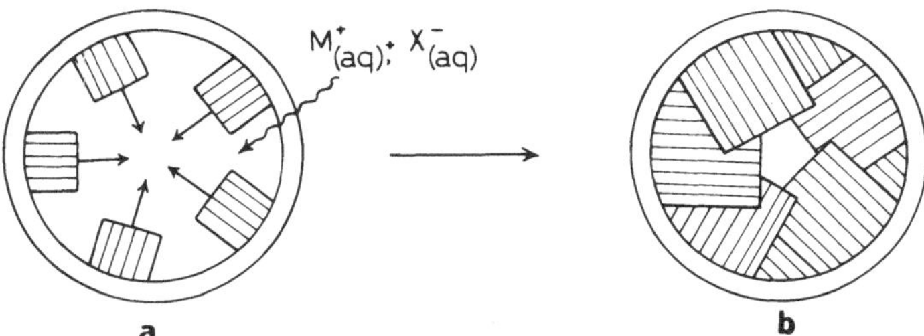

Fig. 4.14 a, b. Epitaxial deposition of minerals within vesicles is limited by the spherical geometry of the compartments; **a** aqueous ions can be transported across the vesicle membrane and precipitated on an atomic net located at the inner membrane surface; **b** epitaxial growth from the membrane sites results in incoherent development of the mineral nuclei and the formation of a polycrystalline material

Elongated vesicles[5,6,9], however, are possible although they will be energetically more demanding, possibly being held in position by cellular microfilaments.

Summarizing, it appears that there is considerable empirical evidence indicating that the role of organic matrices in determining oriented mineral overgrowth in biology may be overestimated and that there are several other possible functions of these frameworks which may be more important. For example, organic matrix-induced processes of oriented growth can not be responsible for the widespread mineralization of SiO_2. There appears to be some *a priori* reasons suggesting that epitaxis, although possible in biomineralization, may not be a fundamental mechanism of oriented growth as often implied, and that other factors such as surface structure and vectorial crystal aggregation may be responsible for biomineral orientation.

5 Amorphous Biominerals

Little attention has been paid to the physical and chemical properties of amorphous phases in biomineralization. The structure of such materials can be regarded to some extent as a highly viscous liquid. The two general models of amorphous structure are, (i) a continuous random network, and (ii) a microcrystalline/cluster network. In the continuous random network model the amorphous state is described in terms of a random network of small structural units which are related to those found in the unit cell of the crystalline state. These units are attached in a random array subject only to the spatial constraints imposed by the steric hindrance of the environment and the allowable ranges for bond angles. In the microcrystalline/cluster model the amorphous structure is described in terms of a random array of microcrystalline polyhedra. Such clusters of icosahedral symmetry may be energetically more favourable when only short range energy contributions are considered. Recent investigations of biogenic silica in choanoflagellates using ultra-high resolution electron microscopy has shown that in this organism the silica is deposited in a continuous random network of SiO_4 tetrahedra (Fig. 5.1)[99]. The structure of biogenic silica can also be studied by solid state ^{29}Si n.m.r. spectroscopy. A preliminary study of plant silica has been reported[100].

Biogenic amorphous minerals may have several functional roles. For example, as precursers for crystalline phases, as ion-stores regenerated in time of biological stress, and as mineralized structures. Transformation of amorphous precursers to crystalline states will be dependent on the activation energy barriers involved and whether such barriers can be modified by interaction of the solid phase with initiators and inhibitors. Thus silica is always deposited under biological conditions in an amorphous hydrated form which is stable due to the high activation energy (*ca.* 200 kcal mol^{-1}) required for transformation into crystalline quartz[42]. Amorphous calcium phosphate, however, readily transforms to more stable crystalline states as previously described. Thus at the onset of biomineralization the mechanism of phosphate and silica deposition may be essentially the same, but with the final mineral phases being dependent on the kinetics of phase-transformation.

Since the solubility of a solid phase increases as the phase becomes thermodynamically more unstable (Table 3.1), amorphous materials, being metastable, will have

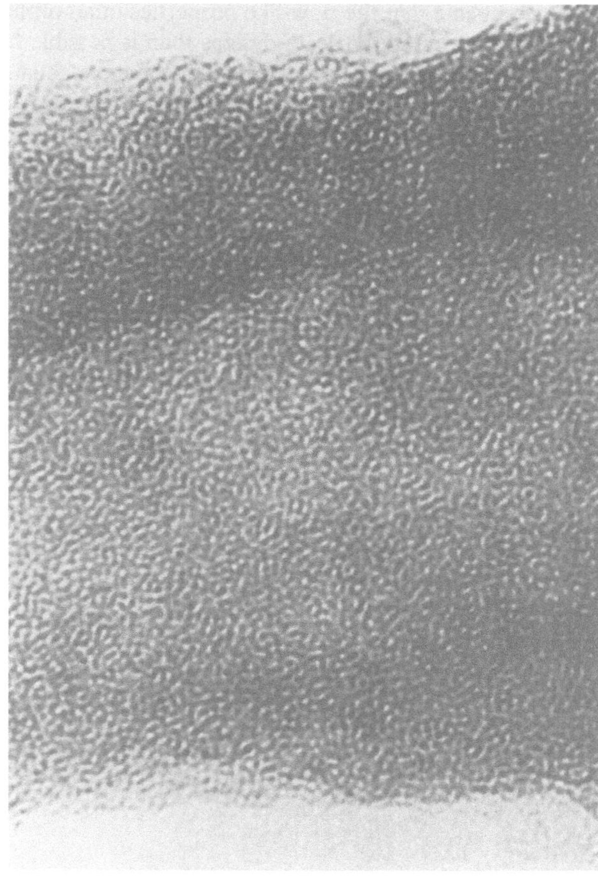

Fig. 5.1. Ultra-high resolution electron micrograph showing the continuous random network of silica in the silicified rods of *Stephanoeca diplocostata* Ellis[99] (× 2,000,000)

greater solubilities than for crystalline states. Thus such materials can be utilised as reservoirs of ions which can, for example, be rapidly regenerated in times of biological stress. The rate of dissolution will be essentially diffusion controlled, and will increase with the degree of hydration of the mineral. Hence the stabilisation of biogenic amorphous minerals requires the presence of an organic sheath such as the silicalemma membrane in diatoms[9].

The use of amorphous minerals as structural elements in biology requires the complete biological control over spatial organization since the material itself has no characteristic morphology. Similarly, there is no possibility of controlling mineral morphology through interaction of the solid surface with growth inhibitors. Thus amorphous materials must be contained within specific geometrically-designed vesicles for them to be stabilised and moulded as part of a skeletal structure. The diversity of diatom structures illustrates the sophistication inherent in these biological design processes[9].

It is possible that the flow-like properties of amorphous materials allows them to be sculptured into more intricate designs than is possible for crystalline biominerals. Thus the gel structure of silica rods formed in choanoflagellates[99)] allows these rods to have a curvature approximating to the radius of curvature of the parent cell. Alternatively, amorphous silica can be deposited in particulate form, for example, in the silica hairs of the weed *Phalaris canariensis*. An interesting characteristic of this system is that at least three types of substructure have been observed in fractured hairs when investigated by electron microscopy[101)]. The alignment of silica into sheet, globular, and fibrillar sub-structures (Fig. 5.2) indicates that there are changes in the spatial organization of the deposited material. One possibility is that the amorphous silica is deposited on an organic framework of changing spatial organization. Deposition on the matrix could occur direct from the contact solution or from the ordered alignment of preformed SiO_2 particles. Alternatively, no matrix may be present, in which case the SiO_2 sub-structures could be determined by changes in the chemical conditions of precipitation (e.g. changes in pH or addition of surfactants[41)], at the mineralizing site.

Fig. 5.2 a–c. Electron micrograph of a fractured silica fibre from the weed *Phalaris canariensis* showing three regions of differing substructure (× 11,000)[101)]. Insets; **a**, sheet (× 56,000); **b**, globular (× 33,000); **c**, fibrillar (× 24,000); arrangements of deposited silica

6 Conclusions

The aim of this paper has been to discuss the solid state factors involved in the regulatory processes of biomineralization. The main conclusions are:

(i) There are three fundamental processes of biological control in mineralization; structural, spatial, and chemical. Crystal chemistry and crystallographic orientation will be determined by the chemistry of the mineralization zone and by the biological design of organic substrate surfaces. Chemical control of the mineralizing site can occur through the maintainence of ion concentration gradients across limiting membranes and by the regulation of accelerator/inhibitor concentrations. Spatial control of biomineralization can result in the regulation of the size and the orientation of mineral particles and can be determined by limiting organic matrices (vesicles and organic frameworks) which act as localised mineralization sites, as surfaces for aggregation of particles of controlled size (vectorial particle aggregation), or as surfaces for oriented growth with crystal size determined by the spatial and temporal regulation of the synthesis of the organic surface. Structural control of biomineralization can be determined by the chemistry of the mineralization zone or by atomic relationships between growing crystal faces and an organic atomic net.

(ii) Although organic matrices may be important as surfaces for oriented growth of biominerals there is much empirical evidence which suggests that this role may be overestimated and that there are several other possible functions of these frameworks, such as surfaces for heterogeneous nucleation and structural support, which may be often more important.

(iii) Oriented growth in biomineralization can occur through three biologically-controlled processes; lattice matching between an organic matrix and depositing crystal faces, surface-structural relationships between an organic matrix and depositing crystal faces, and ordered aggregation of mineral crystallites of controlled size. There appears to be some *a priori* reasons suggesting that epitaxis, although possible in biomineralization, may not be a fundamental mechanism of structural control as often implied, and that non-epitaxial processes may be generally responsible for biomineral orientation.

(iv) Mineral structure will be determined by factors which modulate mineral nucleation and growth. Nucleation can determine the structure of biominerals through interactions between the organic matrix and the depositing ions and by the regulation of the chemical properties of the mineralization zone. The nature of the critical nucleus for nucleation can be an important factor in determining mineral structure. Structural control of biominerals by growth processes can be determined primarily by the physicochemical properties of the mineralization zone, in particular, the rate of phase transformation processes.

(v) Table 6.1 summarises the many possible factors which affect the nucleation, growth, structure, morphology, and size of biogenic minerals. The fundamental concept involved is the modification of the interfacial energies of the growing mineral surface. It seems most probable that an organic surface of some kind (matrix or vesicle membrane) will be the site of nucleation in biologically-controlled mineralization since the control over nucleation in free solution must surely be complicated. The use of vesicles seems to be advantageous in controlling the chemistry of the precipitation process since compartmentalization allows elements to be selected in the proximity of myraid cellular reac-

Table 6.1. Factors affecting the nucleation, growth, structure, morphology and size of biogenic minerals

Nucleation	Growth	Structure	Morphology	Size
supersaturation	supersaturation	supersaturation	supersaturation	supersaturation
nature of critical nucleus	accelerators/ inhibitors	nature of critical nucleus	co-precipitating ions	vesicles
organic substrates	organic nets vesicles	organic nets	organic nets	organic nets
accelerators/ inhibitors	surface defects (dislocations, kinks, steps)	inhibitors of phase transformation	inhibitors	
temperature	temperature	temperature	temperature	temperature
pH	pH	pH	pH	pH

tions. Vesicles are a fundamental component of many silicification processes in which they are utilised in the regulation of chemical and spatial (volume) control.

(vi) Biogenic amorphous mineral phases are important as precursers, ion-stores, and in structural design. They may be macroscopically ordered depending on the chemical and spatial properties within the mineralizing zone.

Acknowledgement. The author is deeply indebted to Professor R. J. P. Williams for many valuable discussions and suggestions in the course of this work.

7 References

1. Lowenstam, H. A.: Science *211,* 1126 (1981)
2. Pautard, F. G. E., Williams, R. J. P.: Chem. in Brit. *18*(3), 14 (1982)
3. Bohm, L., Futterer, D.: J. Phycol. *14,* 486 (1978)
4. Krumbein, W. E.: Biogeochemical cycling of mineral-forming elements (Ed. Trudinger, P. A., Swain, D. J.), Elsevier, Amsterdam 1979
5. Leadbeater, B. S. C.: Protoplasma *98,* 241 (1979)
6. Klaveness, D.: Protistologica *VIII*(3), 335 (1972)
7. Hibberd, D. J.: J. Phycol. *13*(4), 309 (1977)
8. Blakemore, R. P., Frankel, R. B.: Sci. American *245*(6), 42 (1981)
9. Volcani, B. E.: Silicon and siliceous structures in biological systems (Ed. Volcani, B. E., Simpson, T. L.), Springer Verlag 1981
10. Nielson, A. E.: Kinetics of precipitation, Pergamon, Oxford 1964
11. Nielson, A. E., Christoffersen, J.: In Biological mineralization and demineralization (Ed. Nancollas, G. H.), 37–77, Springer-Verlag 1982
12. Walton, A. G.: The formation and properties of precipitates, Interscience, Vol. 23, New York 1967
13. Kossel, W.: Ann. Phys. Chem. *21,* 455 (1934)
14. Stranski, I. N.: Z. Phys. Chem. *A 136,* 259 (1928)
15. Becker, R., Doring, W.: Ann. Phys. *24,* 719 (1935)
16. Becker, R.: Discuss. Faraday Soc. *5,* 50 (1949)

17. Mullin, J. W.: Crystallization, Butterworth & Co., London 1972
18. Stranski, I. N.: Discuss. Faraday Soc. *5,* 13 (1949)
19. Frank, F. C.: ibid. *5,* 48 (1949)
20. Burton, W. K., Cabera, N., Frank, F. C.: Phil. Trans. Roy. Soc. Lond. *A 243,* 299 (1951)
21. Nancollas, G. H.: Adv. Colloid. Interface Sci. *10,* 215 (1979)
22. Doremus, R. H.: J. Phys. Chem. *62,* 1068 (1958)
23. Doremus, R. H.: ibid. *74,* 1405 (1970)
24. Gunn, D. J.: Discuss. Faraday Soc. *61,* 133 (1976)
25. Posner, A. S., Eanes, E. D., Harper, R. A., Zipkin, I.: Arch. Oral Biol. *3,* 549 (1963)
26. Eanes, E. D., Zipkin, I., Harper, R. A., Posner, A. S.: ibid. *10,* 161 (1965)
27. Posner, A. S., Betts, F., Blumenthal, N. C.: Prog. Cryst. Growth Charact. *3,* 49 (1980)
28. Hume-Rothery, W., Mabbott, G. W., Channel-Evans, K. M.: Phil. Trans. Roy. Soc. Lond. *233,* 1 (1934)
29. Berner, R. A.: Geochim. Cosmochim. Acta *39,* 489 (1975)
30. Kinsman, D. J. J., Holland, H. D.: ibid. *33,* 1 (1969)
31. Eanes, E. D., Posner, A. S.: Trans. N.Y. Acad. Sci. *28,* 233 (1965)
32. Eanes, E. D., Meyer, J. L.: Calcif. Tiss. Res. *23,* 259 (1977)
33. Meyer, J. L., Eanes, E. D.: ibid. *25,* 59 (1978)
34. Eanes, E. D., Termine, J. D., Nylen, M. V.: ibid. *12,* 143 (1973)
35. Brown, W. E.: Clin. Orthop. Rel. Res. *44,* 205 (1966)
36. Young, R. A., Brown, W. E.: In Biological mineralization and demineralization (Ed. Nancollas, G. H.), 101–141, Springer-Verlag 1982
37. Hull, H. S., Turnbull, A. G.: Geochim. Cosmochim. Acta *37,* 685 (1973)
38. Turnbull, A. G.: ibid. *37,* 1593 (1973). Quoted value is calculated from this paper; ion-pair contribution not considered
39. Stumm, W., Morgen, J. L.: Aquatic Chemistry, J. Wiley 1981
40. Sillen, L. G., Martell, A. E.: Stability constants of metal-ion complexes, Chem. Soc. 1964
41. Iler, R. H.: Chemistry of silica, J. Wiley 1979
42. Degens, E. T.: Topics Curr. Chem. *64,* 1 (1976)
43. Mann, S., Skarnulis, A. J., Williams, R. J. P.: Isr. J. Chem. *21,* 3 (1981)
44. Hutchison, J. L., Mann, S., Skarnulis, A. J., Williams, R. J. P.: J. Chem. Soc. Chem. Comm. 634 (1980)
45. Mann, S., Williams, R. J. P.: J. Chem. Soc. Dalton Trans. 311 (1983)
46. Mann, S., Ratcliffe, R. G., Kime, M. G., Williams, R. J. P.: ibid. 771 (1983)
47. Mann, S.: D. Phil. Thesis Oxford 1982
48. Watt, F., Grime, G. W., Blower, G. D., Takacs, J., Vaux, D. J. T.: Nuclear Instruments and Methods, North-Holland (in the press) (1982)
49. Bachra, B. N.: Ann. N.Y. Acad. Sci. *109,* 251 (1963)
50. Boskey, A. L., Posner, A. S.: Mat. Res. Bull. *9,* 907 (1974)
51. Fleisch, H., Russell, R. G. G., Bisaz, S., Termine, J. D., Posner, A. S.: Calcif. Tiss. Res. *2,* 49 (1968)
52. Francis, M. D.: ibid. *3,* 151 (1969)
53. Termine, J. D., Conn, K. M.: ibid. *22,* 149 (1976)
54. Howell, D. S., Pita, J. C., Marquez, J. F., Gatter, R. A.: J. Clin. Investig. *48,* 630 (1969)
55. Termine, J. D., Peckauskas, R. A., Posner, A. S.: Arch. Biochem. Biophys. *140,* 318 (1970)
56. Cotmore, J. M., Nichols, G., Wuthier, R. E.: Science *172,* 1339 (1971)
57. Tew, W. P., Mahle, C., Benavides, J., Howard, J. C., Lehninger, A. L.: Biochem. *19,* 1983 (1980)
58. Termine, J. D., Eanes, E. D., Conn, K. M.: Calcif. Tiss. Intern. *31,* 247 (1980)
59. Jackson, T. A., Bischoff, J. L.: J. Geol. *79,* 493 (1971)
60. Levetzow, K. G.: Jena. Z. Naturwiss. *68,* 41 (1932)
61. Birchall, D., Davey, R. J.: J. Cryst. Growth *54,* 323 (1981)
62. Nyvlt, J.: Chem. Prumysl. *12,* 170 (1962)
63. Bryant, G. W., Hallet, J., Mason, B. J.: J. Phys. Chem. Solids *12,* 189 (1959)
64. Wainwright, S. A., Biggs, W. D., Currey, J. D., Gosline, J. M.: Mechanical design in organisms. E. Arnold (1976)
65. Parkes, E. W.: Braced frameworks. Pergamon Press, Oxford 1965

66. Gould, B. S.: Treatise on collagen (3 vols.), Academic Press, New York 1968
67. Bailey, A. J.: The nature of collagen. In Comprehensive Biochemistry, Elsevier Amsterdam, 26-B, 297 (1968)
68. Ramachandran, G. N.: Int. Rev. Conn. Tiss. Res. *1,* 127 (1963)
69. Schmitt, F. O.: Proc. Am. Phil. Soc. *100,* (5), 476 (1956)
70. Smith, J. W.: Nature, Lond., *219,* 157 (1968)
71. Towe, K. M., Lowenstam, H. A.: J. Ultrastruct. Res. *17,* 1 (1967)
72. Coombs, T. L., George, S. G.: Physiology and behaviour of marine organisms (Ed. McLusky, D. S., Berry, A. J.), Pergamon Press, Oxford 1979
73. George, S. G., Pirie, B. J. S.: Biochim. Biophys. Acta *580,* 234 (1979)
74. George, S. G., Pirie, B. J. S.: J. Mar. Biol. Ass. UK. *60,* 579 (1980)
75. Howard, B., Mitchell, P. C. H., Ritchie, A., Simkiss, K., Taylor, M.: Biochem. J. *194,* 507 (1981)
76. Nylen, M. V., Scott, D. B., Mosley, V. M.: Calcification in biological systems (Ed. Sognnaes, R. F.), Washington 1960
77. Anderson, H. C.: J. Cell Biol. *35,* 81 (1967)
78. Bonucci, E.: J. Ultrastruct. Res. *20,* 33 (1967)
79. Eisenmann, D. R., Gluck, P. L.: J. Ultrastruct. Res. *41,* 18 (1972)
80. Gay, C. V., Schraer, H., Hargest, T. E.: Metab. Bone Dis. & Rel. Res. *1,* 105 (1978)
81. Wuthier, R. E., Linder, R. E., Warner, G. P., Gore, S. T., Borg, T. K.: ibid. *1,* 125 (1978)
82. Anderson, H. C.: ibid. *1,* 83 (1978)
83. Salamat, M. S., Ross, M. D., Peacor, D. R.: Ann. Otol. *89,* 229 (1980)
84. Carlstrom, D.: Biol. Bull. *125,* 441 (1963)
85. Nakahara, H., Bevelander, G.: Anat. Rec. *193(2),* 233 (1979)
86. Mann, S., Parker, S. B., Ross, M. D., Skarnulis, A. J., Williams, R. J. P.: Proc. Roy. Soc. Lond. B (in the press) (1983)
87. Weiner, S., Traub, W.: FEBS Letts. *111(2),* 311 (1980)
88. Crenshaw, M. A.: In Biological mineralization and demineralization (Ed. Nancollas, G. H.), 243–257, Springer-Verlag 1982
89. Marsh, R. E., Corey, R. B., Pauling, L.: Biochim. Biophys. Acta *16,* 1 (1955)
90. Van der Merwe, J. H.: Discuss. Faraday Soc. *5,* 201 (1949)
91. Gebhardt, M.: Crystal Growth (Ed. Hartmann, P.), North Holland 1973
92. Finch, G. I., Sun, C. H.: Trans. Faraday Soc. *32,* 852 (1936)
93. Kleber, W.: Growth of crystals *5 A,* 59 (1968)
94. Kleber, W., Ickert, L.: Z. Phy. Chem. *224,* 364 (1963)
95. Bassett, G. A.: Proc. Europ. Reg. Conf. Electron Microscopy *1,* 270 (1961)
96. Hohling, H. J., Kreilos, R., Neubrauer, H. L., Boyde, A.: Z. Zellforsch. *122,* 36 (1971)
97. Jelinski, L. W., Torchia, D. A.: J. Mol. Biol. *133,* 45 (1979)
98. Skarnulis, A. J., Strong, P. J., Williams, R. J. P.: J. Chem. Soc. Chem. Comm. *1030* (1978)
99. Mann, S., Williams, R. J. P.: Proc. Roy. Soc. Lond. B *216,* 137 (1982)
100. Fyfe, C. A., Mann, S., Perry, C. C., Williams, R. J. P., Gobbi, G. G., Kennedy, G. J.: J. Chem. Soc. Chem. Comm. 168 (1983)
101. Mann, S., Parker, S. B., Perry, C. C., Williams, R. J. P.: In Biomineralization and Biological Accumulation 171–183, D. Reidel Publishing Company 1983
102. Williams, R. J. P.: Roy. Soc. Disc., June 1983

Author-Index Volumes 1–54

Reactivity and Structure

Concepts in Organic Chemistry

Editors: K. Hafner, J.-M. Lehn, C. W. Rees, P. v. R. Schleyer, B. M. Trost, R. Zahradník

Volume 1: **J. Tsuji**
Organic Synthesis
by Means of Transition Metal Complexes
A Systematic Approach
1975. 4 tables. IX, 199 pages. ISBN 3-540-07227-6

Volume 2: **K. Fukui**
Theory of Orientation and Stereoselection
1975. 72 figures, 2 tables. VII, 134 pages
ISBN 3-540-07426-0

Volume 3: **H. Kwart, K. King**
d-Orbitals in the Chemistry of Silicon, Phosphorus and Sulfur
1977. 4 figures, 10 tables. VIII, 220 pages
ISBN 3-540-07953-X

Volume 4: **W. P. Weber, G. W. Gokel**
Phase Transfer Catalysis in Organic Synthesis
1977. Out of print. New edition in preparation

Volume 5: **N. D. Epiotis**
Theory of Organic Reactions
1978. 69 figures, 47 tables. XIV, 290 pages
ISBN 3-540-08551-3

Volume 6: **M. L. Bender, M. Komiyama**
Cyclodextrin Chemistry
1978. 14 figures, 37 tables. X, 96 pages
ISBN 3-540-08577-7

Volume 7: **D. I. Davies, M. J. Parrott**
Free Radicals in Organic Synthesis
1978. 1 figures. XII, 169 pages
ISBN 3-540-08723-0

Volume 8: **C. Birr**
Aspects of the Merrifield Peptide Synthesis
1978. 62 figures, 6 tables. VIII, 102 pages
ISBN 3-540-08872-5

Volume 9: **J. R. Blackborow, D. Young**
Metal Vapour Synthesis in Oranometallic Chemistry
1979. 36 figures, 32 tables. XIII, 202 pages
ISBN 3-540-09330-3

Volume 10: **J. Tsuji**
Organic Synthesis with Palladium Compounds
1980. 9 tables. XII, 207 pages. ISBN 3-540-09767-8

Volume 11
New Syntheses with Carbon Monoxide
Editor: **J. Falbe**
With contributions by H. Bahrmann, B. Cornils, C. D. Frohning, A. Mullen
1980. 118 figures, 127 tables. XIV, 465 pages
ISBN 3-540-09674-4

Volume 12: **J. Fabian, H. Hartmann**
Light Absorption of Organic Colorants
Theoretical Treatment and Empirical Rules
1980. 76 figures, 48 tables. VIII, 245 pages
ISBN 3-540-09914-X

Volume 13: **G. W. Gokel, S. H. Korzeniowski**
Macrocyclic Polyether Syntheses
1982. 89 tables. XVIII, 410 pages
ISBN 3-540-11317-7

Volume 14: **W. P. Weber**
Silicon Reagents for Organic Synthesis
1983. XVIII, 430 pages. ISBN 3-540-11675-3

Volume 15: **A. J. Kirby**
The Anomeric Effect and Related Stereoelectronic Effects at Oxygen
1983. 20 figures, 24 tables. VIII, 149 pages
ISBN 3-540-11684-2

Springer-Verlag
Berlin
Heidelberg
New York
Tokyo

Inorganic Chemistry Concepts

Editors: **M. Becke, C. K. Jørgensen, M. F. Lappert, S. J. Lippard, J. L. Margrave, K. Niedenzu, R. W. Parry, H. Yamatera**

Volume 1
R. Reisfeld, C. K. Jørgensen
Lasers and Excited States of Rare Earths
1977. 9 figures, 26 tables. VIII, 226 pages
ISBN 3-540-08324-3

Volume 2
R. L. Carlin, A. J. van Duyneveldt
Magnetic Properties of Transition Metal Compounds
1977. 149 figures, 7 tables. XV, 264 pages
ISBN 3-540-08584-X

Volume 3
P. Gütlich, R. Link, A. Trautwein
Mössbauer Spectroscopy and Transition Metal Chemistry
1978. 160 figures, 19 tables, 1 folding plate. X, 290 pages
ISBN 3-540-08671-4

Volume 4
Y. Saito
Inorganic Molecular Dissymmetry
1979. 107 figures, 28 tables. IX, 167 pages
ISBN 3-540-09176-9

Volume 5
T. Tominaga, E. Tachikawa
Modern Hot-Atom Chemistry and Its Applications
1981. 57 figures, 34 tables. VIII, 154 pages
ISBN 3-540-10715-0

Contents: Introduction. – Experimental Techniques: Production of Energetic Atoms. Radiochemical Separation Techniques. Special Physical Techniques. – Characteristics of Hot Atom Reactions: Gas Phase Hot Atom Reactions. Liquid Phase Hot Atom Reactions. Solid Phase Hot Atom Reactions. – Applications of Hot Atom Chemistry and Related Topics: Applications in Inorganic, Analytical and Geochemistry. Applications in Physical Chemistry. Applications in Biochemistry and Nuclear Medicine. Hot Atom Chemistry in Energy-Related Research. Current Topics Related to Hot Atom Chemistry and Future Scope. – Subject Index.

Volume 6
D. L. Kepert
Inorganic Stereochemistry
1982. 206 figures, 45 tables. XII, 227 pages
ISBN 3-540-10716-9

Contents: Introduction. – Polyhedra. – Four-Coordinate Compounds. – Five-Coordinate Compounds Containing only Unidentate Ligands. – Five-Coordinate Compounds Containing Chelate Groups. – Six-Coordinate Compounds Containing only Unidentate Ligands. – Six Coordinate Compunds $[M(Bidentate)_2(Unidentate)_2]$. – Six-Coordinate Compounds $[M(Bidentate)_3]$. – Six-Coordinate Compounds Containing Tridentate Ligands. – Seven-Coordinate Compounds Containing only Unidentate Ligands. – Seven-Coordinate Compounds Containing Chelate Groups. – Eight-Coordinate Compounds Containing only Unidentate Ligands. – Eight-Coordinate Compounds Containing Chelate Groups. – Nine-Coordinate Compounds. – Ten-Coordinate Compounds. – Twelve-Coordinate Compounds. – References. – Subject Index.

Volume 7
H. Rickert
Electrochemistry of Solids
An Introduction
1982. 95 figures, 23 tables. XII, 240 pages
ISBN 3-540-11116-6

Contents: Introduction. – Disorder in Solids. – Examples of Disorder in Solids. – Thermodynamic Quantities of Quasi-Free Electrons and Electron Defects in Semiconductors. – An Example of Electronic Disorder. Electrons and Electron Defects in α-Ag_2S. – Mobility, Diffusion and Partial Conductivity of Ions and Electrons. – Solid Ionic Conductors, Solid Electrolytes and Solid-Solution Electrodes. – Galvanic Cells with Solid Electrolytes for Thermodynamic Investigations. – Technical Applications of Solid Electrolytes – Solid-State Ionics. – Solid-State Reactions. – Galvanic Cells with Solid Electrolytes for Kinetic Investigations. – Non-Isothermal Systems. Soret Effect, Transport Processes, and Thermopowers. – Author Index. – Subject Index.

Volume 8
M. T. Pope
Heteropoly and Isopoly Oxometalates
1983. 71 figures, 40 tables. XIII, 180 pages
ISBN 3-540-11889-6

Contents: Introduction. – Preparation, Structural Principles, Properties and Applications. – Isopolyanions. – Heteropolyanions. – Heteropolyanions and Ligands. – Redox Chemistry and Heteropoly Blues. Organic and Organometallic Derivatives. – Polyoxometalate Chemistry. Current Limits and Remaining Challenges. – Appendix: Nomenclature of Polyanions. – References. – Subject Index.

Springer-Verlag
Berlin
Heidelberg
New York
Tokyo